AIR FILTRATION

An Integrated Approach to the
Theory and Applications of Fibrous Filters

Pergamon Titles of Related Interest

BANHIDI
Radiant Heating Systems

DODGSON
Inhaled Particles VII

MASUDA & TAKAHASHI
Aerosols: Science, Industry, Health and the Environment

VINCENT
Ventilation 1988

VISUALIZATION SOCIETY OF JAPAN
Atlas of Visualization

WILLIAMS & LOYALKA
Aerosol Science: Theory and Practice

Pergamon Related Journals

Annals of Occupational Hygiene

Atmospheric Environment: Part A: General Topics,
Part B: Urban Atmosphere

Building and Environment

Current Advances in Ecological and Environmental Science

Environment International

International Journal of Heat and Mass Transfer

International Journal of Multiphase Flow

Journal of Aerosol Science

Free sample copies of journals are gladly sent on request

AIR FILTRATION

An Integrated Approach to the Theory and Applications of Fibrous Filters

R. C. BROWN

Health and Safety Executive
Research and Laboratory Services Division,
Sheffield, UK

PERGAMON PRESS

OXFORD · NEW YORK · SEOUL · TOKYO

UK	Pergamon Press, Ltd, Headington Hill Hall, Oxford OX3 0BW, England
USA	Pergamon Press Inc., 660 White Plains Road, Tarrytown, New York 10591-5153, USA
KOREA	Pergamon Press Korea, KPO Box 315, Seoul 110-603, Korea
JAPAN	Pergamon Press Japan, Tsunashima Building Annex, 3-20-12 Yushima, Bunkyo-ku, Tokyo 113, Japan

Copyright © 1993 R. C. Brown

First edition 1993

British Library Cataloguing in Publication Data
A catalogue record for this book is available from the British Library.

Library of Congress Cataloging in Publication Data
Brown, R. C. (Richard Colin), 1944–
Air filtration: an integrated approach to the theory and applications of fibrous filters/R.C. Brown.—1st ed.
p.   cm.
Includes bibliographic references.
1. Air filters.   2. Fibers.   3. Air–Purification.   I. Title.
TH7683.A3B76   1993
628.5'.3—dc20

ISBN 0 08 041274 2

Transferred to digital printing 2005

Contents

$E_{\sigma q}$ Single fibre efficiency by line-dipole fibre with charged particles

E_{Oq} Single fibre efficiency by neutral fibre with charged particles

E_{ROq} Single fibre efficiency by neutral fibre with charged particles, and interception

E_{pq} Single fibre efficiency by polarised fibre with charged particles

E_{pO} Single fibre efficiency by polarised fibre with neutral particles

$f(c)$ Dimensionless function relating pressure drop to packing fraction

F Force acting on a particle

F_d Drag force acting on unit length of fibre

F_s Force per unit area caused by viscous drag

$g(r)$ Radial component of stream function

$G_1(r)$ Free space solution of Laplace's equation, see equation 3.55

$G_2(r)$ Free space solution of biharmonic equation, see equation 3.56

h Filter thickness

hv Lifshitz–van der Waals constant

I Probability of intersection of fibres in a square of side equal to the fibre length

$\underline{k}$ Tensor relating pressure gradient to air velocity in a filter

k As above when scalar in form, i.e. medium homogeneous and isotropic

k_B Boltzmann's constant

Kn Knudsen number

Ku Kuwabara constant, equal to $-\frac{1}{2}\ln(c)-0.75+c+c^2/4$

K_0 Modified Bessel function of zero order

K_1 Modified Bessel function of first order

K Shape factor of fibre

$K_{\parallel}$ Component of shape factor for airflow in direction of fibre axis

$K_{\perp}$ Component of shape factor for airflow perpendicular to direction of fibre axis

L Length of fibre per unit volume of filter

m Particle mass

m_p Dipole moment per unit length of fibre polarised by electric field E

m_k Number of particles in kth layer of dendrite

m_{tot} Total number of particles captured by a fibre

M Quantity of aerosol deposited in unit volume of filter

Introduction

THE history of filtration (1) has, in some way, reflected the history of human needs. At the present time the necessity for clean air exists along with the necessity for a large number of industrial processes; and certain parts of industry itself rely on clean air of exceptional quality. Many employees in industry or agriculture need to avoid exposure to airborne particulates, and protection is usually provided by respirator filters or larger scale filtration units. The medical and biological fields rely on sterile air; and filtration can also improve air quality in the office, the home and in transport of all types. Industrial hygienists and engineers require an understanding of filtration, as does anyone whose occupation involves the maintenance or monitoring of air quality.

Because the action of the simplest type of filter, a net, is elementary, it might be thought that this simplicity extends to filters proper. In practice filters act in a variety of subtle ways, and the mechanism of action of a single filter varies with conditions. Filtration is a subject that draws on fluid mechanics, aerosol science, electrostatics and the science/engineering interface.

A study of filtration can be made at a variety of levels, and I have tried to write the book with this in mind. Where possible, each chapter or principal section contains paragraphs explaining its subject matter in a simple way and deriving formulae from first principles, even if this requires rather severe approximations. The rest of the text tackles the problem in greater depth. I hope that this will make the book useful to anyone needing only a superficial understanding; but I have found myself that a simplified explanation of scientific principles can often serve as a good basis for detailed study.

The depth to which the subject is covered varies throughout the book, depending on the extent to which research has been carried out, and data collected and analysed. Studies of airflow and of the capture of spherical particles by mechanical means have reached quite an advanced level, but both theory and experiment on non-spherical particles are sparse. In Chapter 7 experimental data on the adhesion and bounce of particles has been supplemented by data obtained using planar substrates, because these will illustrate aspects of behaviour that do not depend on geometry. Similarly, in Chapter 8, some of the theory of clogging comes from work done on spherical collectors, because these are easier to describe, and they illustrate certain fundamental points. The complexity of these two parts of the subject has meant that not all theories are consistent; and the high dependence of results on experimental conditions has meant that not all observers are in agreement.

The relatively large number of equations in the text may, at first sight, be off-putting to the reader who requires little mathematical involvement, but in most instances the equations are of a simple kind. A large number of physical properties requires a large number of symbols, and I have tried to use those employed by workers in the field, or to replace them by familiar forms. Sometimes duplication has remained, and it has frequently made subscripting necessary.

The subject has a large and ever-increasing bibliography, and a problem faced by an author is not what to include but what to leave out. In general I have not discussed in detail work that deals purely with mathematical models of filter behaviour, unless the work points to the description of observable quantities; though I have tried to include references where possible. Likewise I have tended not to quote experimental work that refers only to highly specialised filters. Parts of the subject are just being opened to study, others are on the fringes of the field, and in these cases I have covered matters thinly, using a sentence and references.

I have referred to the work of a large number of authors, sometimes in detail but often with considerable abbreviation. Where possible I have tried to present a simplified treatment of the original work whilst attempting to do the contributor justice. If I have failed I apologise.

I am indebted to the large number of people who have collaborated with me on filtration, or who have helped me directly or indirectly in the course of my work. There are too many for all to be mentioned personally but I would particularly like to thank two former colleagues at the Health and Safety Executive: Mr G. K. Greenough, who introduced me to the study of filtration, and Dr J. G. Firth who first suggested that I should write this book. I would also like to thank Dr J. I. T. Stenhouse of Loughborough University for reading part of the text. Most of all I am grateful to the numerous authors whose work I have cited.

Reference

1. DAVIES, C.N. Fibrous filters for dust and smoke. *Proceedings of the 9th International Congress on Industrial Medicine, Simpkin Marshall, London,* John Wright, Bristol, 1949.

Symbolic Notation

$A(\alpha)$	Distribution of layer efficiencies
A_1	Hamaker constant
b	Half-distance between parallel fibres in uniform layer
c	Packing fraction
Cn	Cunningham slip correction factor
d_f	Filter fibre diameter
d_p	Particle diameter
d_s	Particle stopping distance
D	Coefficient of diffusion
d_e	Minor axis of prolate ellipsoid, or fibre diameter
d_m	Mass median particle diameter
d_n	Number median particle diameter
D_p	Dielectric contant of material of particle
D_f	Dielectric constant of material of fibre
e	Half-layer spacing in fibre array
e_r	Coefficient of restitution
e_{pl}	Coefficient of restitution of plastic deformation
e_0	Geometry independent fraction of coefficient of restitution
$e'(N_R)$	Contribution to coefficient of restitution defined in equation 7.10
$\mathbf{E}$	Electric field
E_s	General single fibre efficiency
E_R	Single fibre efficiency by interception
E_G	Single fibre efficiency by gravity
E_D	Single fibre efficiency by diffusional deposition
E_I	Single fibre efficiency by inertial impaction
E_{12}	Single fibre efficiency by general processes 1 and 2
E_{GR}	Single fibre efficiency by gravity and interception
E_{QO}	Single fibre efficiency by charged fibre with neutral particles
E_{Qq}	Single fibre efficiency by charged fibre with charged particles
$E_{\sigma O}$	Single fibre efficiency by line-dipole fibre with neutral particles

$E_{\sigma q}$	Single fibre efficiency by line-dipole fibre with charged particles
E_{Oq}	Single fibre efficiency by neutral fibre with charged particles
E_{ROq}	Single fibre efficiency by neutral fibre with charged particles, and interception
E_{pq}	Single fibre efficiency by polarised fibre with charged particles
E_{pO}	Single fibre efficiency by polarised fibre with neutral particles
$f(c)$	Dimensionless function relating pressure drop to packing fraction
F	Force acting on a particle
F_d	Drag force acting on unit length of fibre
F_s	Force per unit area caused by viscous drag
$g(r)$	Radial component of stream function
$G_1(r)$	Free space solution of Laplace's equation, see equation 3.55
$G_2(r)$	Free space solution of biharmonic equation, see equation 3.56
h	Filter thickness
hv	Lifshitz–van der Waals constant
I	Probability of intersection of fibres in a square of side equal to the fibre length
$\underline{k}$	Tensor relating pressure gradient to air velocity in a filter
$\overline{k}$	As above when scalar in form, i.e. medium homogeneous and isotropic
k_B	Boltzmann's constant
Kn	Knudsen number
Ku	Kuwabara constant, equal to $-\frac{1}{2}\ln(c) - 0.75 + c + c^2/4$
K_0	Modified Bessel function of zero order
K_1	Modified Bessel function of first order
K	Shape factor of fibre
$K_\parallel$	Component of shape factor for airflow in direction of fibre axis
$K_\perp$	Component of shape factor for airflow perpendicular to direction of fibre axis
L	Length of fibre per unit volume of filter
m	Particle mass
m_p	Dipole moment per unit length of fibre polarised by electric field **E**
m_k	Number of particles in kth layer of dendrite
m_{tot}	Total number of particles captured by a fibre
M	Quantity of aerosol deposited in unit volume of filter

M_L	Mass load per unit area of filter
M_0	Mass load that doubles single fibre efficiency
n	Number of particles captured per unit length of fibre per unit time
n_q	Number of elementary charges held by a particle
n_s	Number of particles in a latex droplet
N	Aerosol concentration
N_f	Number of fibres per unit area of (paper) filter
N_P	Number of pores per unit area of filter
$N_I(t)$	Ionic concentration after time t
N_R	Dimensionless parameter describing capture by interception
N_G	Dimensionless parameter describing capture by gravity
N_t	Number of particles counted in unit time
N_{Qq}	Dimensionless parameter describing capture of charged particles by a charged fibre
N_{QO}	Dimensionless parameter describing capture of neutral particles by a charged fibre
$N_{\sigma Q}$	Dimensionless parameter describing capture of charged particles by a line-dipole charged fibre
$N_{\sigma O}$	Dimensionless parameter describing capture of neutral particles by a line-dipole charged fibre
N_{Oq}	Dimensionless parameter describing capture of charged particles by a neutral fibre
N_{pq}	Dimensionless parameter describing capture of charged particles by a polarised fibre
N_{pO}	Dimensionless parameter describing capture of neutral particles by a polarised fibre
p	Pressure
p_{p1}	Microscopic yield pressure
P	Aerosol penetration
$P_N(r)$	Probability function of nearest neighbour distance, r
Pe	Peclet number
$P_\tau(t)$	Probability of finding a charge carrier in a state of lifetime τ after a time, t
q	Charge held by a particle
Q	Charge per unit length of fibre
QF	Quality factor
R	Fibre radius
R_a	Surface asperity radius
R_c	Recombination coefficient
Re	Reynolds number
$S(d_p)$	General normalised particle size distribution
St	Stokes number

t	Time
t_o	Total experimental time
T	Absolute temperature
U	Fluid (air) velocity
U_x, U_y	Cartesian components of U
U_r, U_θ	Polar components of U
V	Particle velocity
V_i	Particle approach velocity
V_o	Particle departure velocity
V_c	Particle critical velocity for adhesion
V_d	Particle drift velocity
V	Electrostatic potential
w	Complex cartesian coordinate, $\pi(x+iy)/2b$
W	Complex fluid velocity, $U_x - iU_y$
$W(d)$	Weibull distribution
x	Cartesian coordinate
y	Cartesian coordinate
z	Cartesian coordinate
z_o	Distance of closest approach during adhesion
α	Layer efficiency or filtration index
β	Aspect ratio of fibre
β_{TL}	Degradation parameter
γ	Euler's constant
γ_g	Constant in gamma distribution
γ_i	Lagrangian un-determined multiplier
Γ	Velocity gradient
$\Gamma(\gamma)$	Gamma function
δ	Distance from fibre surface in normal direction, $r - R$
δ_W	Constant in Weibull distribution
Δp	Pressure drop across a filter
ε	Quotient of filter pressure drop and that of a fan-model filter with the same fibre diameter and packing fraction
ε_0	Permittivity of free space
ζ	Hydrodynamic factor
η	Coefficient of viscosity
η_W	Constant in Weibull distribution
θ	Angular polar coordinate
θ_G	Angle between gravity and flow direction
θ_g	Constant in gamma distribution
Θ	Rate of dissipation of energy
λ	Mean free path of air molecules
μ_s	Mean number of particles in a latex suspension droplet
μ	Particle mobility
μ_e	Particle electrical mobility

μ_ω	Particle angular mobility
μ_F	Mean number of fibres per unit area of sectioned fibres
μ_m	Mass mean particle diameter
μ_n	Number mean particle diameter
$v(r)$	Function defined in boundary element theory
ζ	Parameter relating to aerodynamic slip, defined in equation 3.61
ρ	Particle material density
ρ_g	Air density
ρ_m	Maximum number of particles in dendrite layer that can adhere to one particle in supporting layer
σ	Surface charge density of fibre
σ_g	Geometric standard deviation
σ_α	Standard deviation of layer efficiency
σ_c	Standard deviation of filter packing fraction for non-uniform filter
σ_{rr}	Stress tensor component giving stress in r-direction acting on a surface with its normal in the r-direction
$\sigma_{r\theta}$	Stress tensor component giving stress in θ-direction acting on a surface with its normal in the r-direction
τ_s	Particle stopping time in Stokes flow
τ_T	Surface tension
τ	Charged state lifetime
Φ	State binding energy
Φ_{VW}	Van der Waals interaction energy
Φ_{LJ}	Lennard-Jones interaction energy
ϕ	Three-dimensional stream function
ϕ_k	Rate constant for dendrite formation
χ	Constant in Spielman Goren theory $= \sqrt{kR}$
ψ	Stream function
ω	Vorticity

CHAPTER 1

Macroscopic Behaviour of Filters

Surface filtration and depth filtration

On first coming across a fibrous filter one might consider its behaviour to be similar to that of a net or a sieve; but though filters and nets have the same basic purpose their methods of action are different. A net will be 100% efficient in the capture of particles that are larger than its perforations, and the captured objects will be in contact with a substantial part of the structure of the net. The size of the holes is critical, but the thickness of the net is not. Two identical nets in series will not perform better than one, which means that the performance can be understood by considering the surface alone; the process is surface filtration. On the other hand, thick fibrous filters are more efficient than thin ones, and no fibrous filter is 100% efficient.

A useful way to think of a filter is as a large number of layers, each sparsely populated with fibres. Even if a single layer has a very low capture efficiency, the filter as a whole can perform well. For example, if a filter is made up of one hundred fibre layers, each of which captures only 5% of the incident particles, the filter as a whole will capture 99.4%. This process, particle capture throughout the filter, is termed depth-filtration; and it is depth-filtration in fibrous filters that forms the greater part of the following chapters.

Fibrous filters are pads of fibres in an open three-dimensional network. Points of fibre–fibre contact are relatively infrequent, and fibrous filters can have considerable rigidity even though the packing fraction (the fraction of the perceived volume of the filter that is actually occupied by the fibres) is only a few percent. Because of their open structure, fibrous filters are highly permeable to the air and they offer a low resistance to airflow. An electron micrograph of a typical fibrous filter is shown in Fig. 1.1.

Depth filters are able to capture particles that are far too small to be sieved out. Figure 1.2 shows that particles with diameters of only a few

Fɪɢ. 1.1.   Electron micrograph of fibrous filter, illustrating basic fibre alignment and crimp.

micrometres are efficiently captured by a filter with fibres approximately 20 μm in diameter and a packing fraction of 0.05, even though the interfibre spaces of the filter are up to 100 μm in size. Furthermore, the filter, shown in Fig. 1.3, is so open in structure that a captured particle of even 10 μm in diameter would be very unlikely to be in simultaneous contact with more than one fibre. Particle capture must, therefore, involve just one fibre; and for this reason, the theory of particle capture leans heavily on single fibre theory, which will be discussed in detail later.

Types of filter

Apart from fibrous filters there exist granular filters, fabric filters, and membrane filters. Granular filters, which consist of packed beds of roughly isometric particles and which act largely by depth filtration, are treated in detail elsewhere (1), and will not be discussed at length here.

Fabric filters are made from textile fibres, which are processed into a relatively compacted form by weaving or felting. Although some depth filtration does occur in fabric filters, most of the dust does not penetrate into the material but is captured on the surface as a cake (2). Fabric

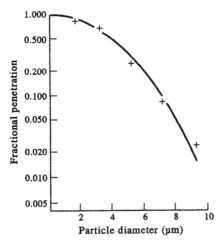

Fig. 1.2   Penetration of monodisperse particles through simple filter as a function of particle size, illustrating low capture efficiency of small particles. (© Crown Copyright; by permission of Her Majesty's Stationery Office.)

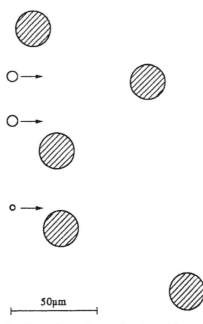

Fig. 1.3   Section of a filter illustrating scale of particles and fibres. (© Crown Copyright; by permission of Her Majesty's Stationery Office.)

filters are usually cleanable, but their resistance to airflow is relatively high. Fibrous filters usually cannot be cleaned and must be disposed of when spent, but their resistance is low.

Membrane filters are made from perforated material or highly compacted fibrous material, usually only a few micrometres thick (3) and acting principally by surface filtration. Detailed study of flow patterns has tended to be limited to the behaviour of the air upstream of the filter surface (4, 5), but extensive studies of their performance have been made (6, 7).

Finally, particulate filters like all of the above must be distinguished from filters used to remove gases or vapours from air by adsorption (8). The efficiency of adsorptive filters depends on the kinetics of adsorption of the molecules of gas or vapour, for the high diffusivity of molecules ensures that they readily come into contact with the material of the filter. The basic technical problem of particulate filtration, on the other hand, is to bring the particles of suspended material into contact with the collectors.

Method of filtration and assessment of efficiency

Both the type of filter used and the method by which the efficiency of filtration should be assessed depend on the purpose of the filter and on what is to be done with the captured material. Airborne particulates are often produced during the processing of materials of value, in which case the substances removed from the air during filtration will be collected and the air carrying them discharged to the atmosphere. The captured material might be a pollutant removed from a measured quantity of air so that it can be identified and quantified. The filter could be used for air or gas cleaning, in which case it is the filtrate that is of value; the particulate material may be a noxious substance removed to make the air fit to be breathed, or it may be non-toxic matter removed to provide air of exceptional quality, necessary for certain manufacturing processes.

In the first case recovery of material is of paramount importance, and so filtration in which a deposit is collected on the surface of the filter, and which can be easily separated from it, will probably be favoured. In the second case, the material should be easily distinguishable from the filter, and so again surface filtration is likely to be used. In the last case, a contaminated filter may well be disposed of along with the material that it contains, and for this reason the use of depth filters may be convenient.

In material recovery or collection for analysis, system performance should be assessed in terms of the quantity of material collected; for example a collection efficiency of 99% does not differ greatly from one of 99.9%. If, on the other hand, one is considering air quality, the 1% penetration in the first instance is an order of magnitude worse than the 0.1% penetration in the second. For this reason the performance of filters used for air cleaning, respiratory protection or any other purpose in which it is the quality of the filtered air or gas that is of interest, should be quantified in terms of aerosol penetration.

Layer efficiency

The concept of a depth filter being made up of a large number of layers, introduced above, gave a useful description of its behaviour. This idea can be extended to quantify the efficiency of filters in the capture of aerosols challenging them.

Layer efficiency for monodisperse aerosols

An aerosol made up of particles of a single size is termed monodisperse; one with a range of particle sizes is called polydisperse. These terms are awkward, but aerosol scientists are stuck with them. The penetration of identical particles through a homogeneous filter that captures particles throughout its depth is related to the thickness. The number captured by a layer of thickness δx will be proportional to the number incident, to the thickness δx and to a constant describing the efficiency of the filter.

$$\delta N = -\alpha N \, \delta x \qquad (1.1)$$

In limiting conditions equation 1 becomes a simple differential equation with the solution,

$$N(x) = N(o)\exp(-\alpha x) \qquad (1.2)$$

where α, which has dimensions L^{-1}, is the layer efficiency or the filtration index. Manipulation of this equation gives,

$$\ln(P) = \ln\left(\frac{N(x)}{N(o)}\right) = -\alpha x. \qquad (1.3)$$

which means that the penetration of the aerosol through the filter, plotted on a logarithmic scale against thickness on a linear scale gives a straight line, the gradient of which is directly proportional to the layer efficiency of the filter material.

Such a plot is shown as graph 1 in Fig. 1.4, which is the penetration/depth curve for a filter/aerosol system with a layer efficiency of 0.1 mm^{-1} (layer efficiency is as much a property of the aerosol as it is of the filter).

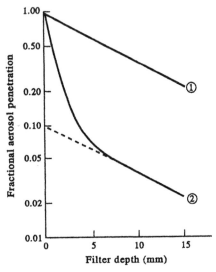

FIG. 1.4   Illustration of layer efficiency: (1)   monodisperse aerosol with $\alpha =$ 0.1 mm^{-1}; (2)   two-component aerosol with $\alpha_1 = 0.1$ mm^{-1}, $\alpha_2 = 1.0$ mm^{-1}, $\beta =$ 0.9.

Equation 1.1 can be generalised slightly by including the quantity of aerosol captured per unit volume of filter material, M, giving a mass balance equation (9),

$$-\left.\frac{\partial N}{\partial x}\right|_t U = \left.\frac{\partial M}{\partial T}\right|_x \tag{1.4}$$

where U is the velocity at which the aerosol approaches the filter. Equation 1.4 is of particular use when the mass of deposited material becomes important, as in filter clogging, to be treated in Chapter 8.

Monodisperse aerosols are a useful experimental tool in the laboratory, though they rarely occur naturally, with the exception of certain biological aerosols (10). Nevertheless, the simple form of Fig. 1.4, with ordinates plotted on a logarithmic scale and abscissae on a linear scale, is a clear way of displaying data for aerosols that are not monodisperse.

It is often useful to plot penetration against some other property of the filter such as particle size, and in such instances the log/linear plot is still likely to be the most fundamental and informative way of displaying the data.

Layer efficiency for polydisperse aerosols

Equation 1.3 can be generalised to describe what happens if a polydisperse aerosol, one containing particles with a range of layer efficiencies, is passed through the filter. Let us consider for the start an aerosol with only two types of particle, the first having a layer efficiency α_1, and the second α_2, with the second particles making up a fraction, β, of the total aerosol. The penetration of this aerosol is,

$$P = (1 - \beta)\exp(-\alpha_1 x) + \beta \exp(-\alpha_2 x) \tag{1.5}$$

and a plot of the penetration as a function of depth is shown as curve 2 in Fig. 1.4, in which α_1 and α_2 are 0.1 and 1.0 mm^{-1} respectively, and β is 0.9. The graph is concave upwards. For small values of x the graph falls steeply, because most of the particles captured will be the α_2 particles, which make up 90% of the aerosol. For large values of x the curve has, as an asymptote, a line of gradient $-\alpha_1$, because these will be virtually the only particles left. Moreover the intercept of this asymptote will give the fraction of particles of this type, as Fig. 1.4 shows.

If we now generalise the description to a polydisperse aerosol with a continuous distribution of layer efficiencies, such that a fraction $A(\alpha)$ has layer efficiencies, α, the expression for penetration becomes,

$$P(x) = \int_0^\infty A(\alpha)\exp(-\alpha x)\, d\alpha. \tag{1.6}$$

In equation 1.6 $P(x)$ is the Laplace transform of $A(\alpha)$; and some interesting properties of $P(x)$ follow from general arguments. The gradient of the logarithm of penetration with respect to x is always negative, and the second derivative is always non-negative. This can be proved mathematically, and an example is that shown in Fig. 1.5, the positive second derivative meaning that the curve is concave upwards. The least penetrating particles are captured early in the passage of the aerosol through the filter, with the result that the remaining aerosol is relatively depleted of easily captured particles and is therefore more penetrating.

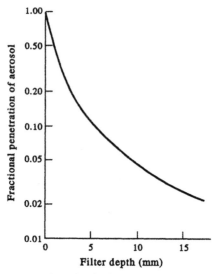

FIG. 1.5   Penetration of polydisperse aerosol through a filter.

The method of moments can be used to show that whatever the form of $A(\alpha)$, the gradient at $x=0$ is the same as that of a monodisperse aerosol with a layer efficiency given by the arithmetical mean of $A(\alpha)$, and the second derivative at $x=0$ is related to the mean layer efficiency and the standard deviation, σ_α,

$$\left.\frac{\partial^2 P}{\partial x^2}\right|_{x=0} = \sigma_\alpha^2 + \bar{\alpha}^2. \tag{1.7}$$

Both the mean layer efficiency of the aerosol and its standard deviation decrease as it passes through the filter. Consider, for example, an aerosol consisting of a simple mixture of equal numbers of particles, half with layer efficiency α and half with layer efficiency 2α. The mean layer efficiency will be $3\alpha/2$ and the standard deviation $\alpha/2$. A filter that removes half of the first set of particles will remove three quarters of the second set. The filtered aerosol will have a mean layer efficiency of 1.33α and a standard deviation of 0.47α. It has become more penetrating and less disperse.

It will be shown in Chapter 4 that there exists a "most-penetrating" type of particle in any aerosol. This means that α will have a non-zero lower limit, α_{min}. As the aerosol passes through the filter, becoming more and more depleted of particles, its behaviour will become increasingly close to that of a monodisperse aerosol with layer efficiency α_{min}.

Quality factor

The pressure drop across a filter is proportional to both its thickness and its area weight, provided that the filtration velocity is constant and the filter homogeneous, and so equations similar to equations 1.2 and 1.3 can be written down and graphs similar to Figs 1.4 and 1.5 can be plotted for both pressure drop and area weight. Since pressure drop is related to energy expenditure in filtration, the quotient of the logarithm of the penetration and the pressure drop is a measure of the performance achieved against the energy expended. This quotient is called the quality factor QF;

$$QF = \frac{-\ln(P)}{\Delta p}. \tag{1.8}$$

Alternative definitions of quality factor (11) contain numerical constants; but however quality factor is defined, it is a function of particle size and filtration velocity, but not of filter thickness.

It is a useful quantity so far as monodisperse aerosols are concerned, but the application of equation 1.8 to experimental data for polydisperse aerosols can give misleading results. In the case of the two-component aerosol, a measurement made on a thin filter will give a value close to that appropriate for the more efficiently captured component, and a measurement on a thick filter will give a value appropriate to the other component. In the case of any polydisperse aerosol the quality factor measured will decrease steadily for thicker filters. This is important to note because not only does it mean that measurements of quality factor are ambiguous, it also means that in practice an effort to improve efficiency by increasing the filter thickness gives diminishing returns.

Single fibre efficiency

Up to now no detailed account has been taken of filter structure; the theory has been macroscopic. A detailed account of structure and structure-related behaviour is given in the following chapters; and a link between the two approaches is needed, which takes the form of the relationship between layer efficiency and single-fibre efficiency.

If a fibre in a filter is orientated at right angles to the flow, the area that it presents is equal to the product of the length and the diameter of the fibre. A fibre that has an efficiency of unity removes from the air all of the particles that would lie within the volume swept out by its area and the velocity vector of the air, assumed to be flowing uniformly, as illustrated in Fig. 1.6. In general a fibre does not remove all of these

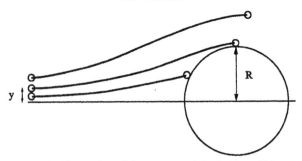

FIG. 1.6 Illustration of the concept of single fibre efficiency.

particles; and the single fibre efficiency is defined as the quotient of the number of particles actually removed and the number that would be removed by a 100% efficient fibre. Its value, y/R in Fig. 1.6, is usually much less than unity, but it can exceed unity in certain cases. Single fibre efficiency, E_s, which is dimensionless, can be related to layer efficiency in the following way,

$$E_s = \frac{\alpha\pi \, d_f}{4c}. \tag{1.9}$$

Equation 1.9 can be derived using simple geometry, if it is assumed that all of the fibres lie perpendicular to the airflow. It depends on the total length of fibre in unit volume of filter, which is related to the packing fraction, c.

Some authors (12) quote a slightly different relationship between layer efficiency and single fibre efficiency.

$$E_s = \frac{\alpha\pi \, d_f(1-c)}{4c} \tag{1.10}$$

Both this and equation 1.9 are correct, provided that E_s is defined in the manner appropriate to each. The discrepancy arises because the fraction, c, of the filter volume occupied by the fibres is unavailable for airflow, and so the mean velocity of the air within the filter is higher by a factor $(1-c)^{-1}$ than the velocity of the air approaching the filter.

This can be considered to apply to the mean velocity of the airborne particles; and if the distance, y, in Fig. 1.6 is measured at a point where the mean particle velocity equals that of the approaching air, equation 1.9 holds. If it is measured at a point where the mean particle velocity equals that of the interstitial air, equation 1.10 applies. The latter definition is more logical, though conditions of uniform particle velocity are unlikely to be realised within the filter itself.

Alternatively, the two definitions of single fibre efficiency refer to the fraction of aerosol captured relative to that swept out by the fibre from air at the velocity approaching the filter, and from air at the (higher) mean velocity within the filter. In Chapter 4 more general expressions for single fibre efficiency are given, along with conditions necessary for the concept to be valid, but equations 1.9 and 1.10 form the basic connection between a macroscopic and a microscopic theory of filtration.

References

1. TIEN, C. *Granular Filtration of Aerosols and Hydrosols*, Butterworths, Boston, 1989.
2. SVAROVSKY, L. *Solid–Gas Separation*, Elsevier, Amsterdam, 1981.
3. BROCK, T. D. *Membrane Filtration: A User's Guide and Reference Manual*, Springer-Verlag, Berlin, 1983.
4. MANTON, M. J. The impaction of aerosols on a nuclepore filter, *Atmos. Environ.*, 1978, **12**, 1669–1675.
5. BROWN, R. C. Viscous flow into a membrane filter composed of randomly situated pores, *J. Aerosol Sci.*, 1983, **14**(4), 481–489.
6. SPURNY, K. Nuclepore Siebfilter Membranen: Zehn Jahre anwendug fur Staub und Aerosolmessungen, *Staub Reinhalt. Luft.*, 1977, **37**(9), 328–334.
7. GENTRY, J. W. and SPURNY, K. R. Measurements of the collection efficiency of nuclepore filters for asbestos fibers, *J. Coll. Int. Sci.*, 1978, **65**(1), 174–180.
8. SCHWEITZER, P. A. *Handbook of Separation Techniques for Chemical Engineers*, McGraw Hill, New York, 1979.
9. DULLIEN, F. A. I. Maximizing the capacity and life of depth-type filters, *Can. J. Chem. Eng.*, 1989, 689–692.
10. HUMPHREY, A. E. and GADEN, E. L. Air sterilization by fibrous filter media, *Ind. Eng. Chem.*, 1955, **47**, 924–930.
11. DORMAN, R. G. In *Aerosol Science* (ed. C. N. DAVIES), Academic Press, London, 1966.
12. PICH, J. In *Aerosol Science* (ed. C. N. DAVIES), Academic Press, London, 1966.

CHAPTER 2

Filter Structure

Introduction

The structure of filters, perhaps the most fundamental of their properties, has received comparatively little attention. Much theoretical work or interpretation of experimental results starts with the assumption that a particular simplified description of structure embodies the essential features of the filter. This sort of assumption is necessary if we are to reduce the problems to a tractable form, but our knowledge of filtration is far from complete without at least a descriptive understanding of real structure.

The structure of granular filters is much easier to describe than that of fibrous filters; because the fundamental units are relatively isometric, and if the nature of the packing is understood, the structure will be known. If the granules were arranged in the simple-cubic, body-centred cubic, face-centred cubic or hexagonal-close-packed configurations encountered in crystallography (1), the resulting structure would be porous. However, if filter fibres are assumed to be cylinders, the only idealised packing, that in which the fibres are aligned and packed as a bundle, would allow the transport of air in one direction only and would not produce an effective filter. Simple packing models could, therefore, be reasonable approximations to the structure of granular filters but not to that of fibrous filters.

Anyone who picks up a particulate filter capsule (a fibrous filter) and a vapour filter capsule (a granular filter) will immediately notice that the latter is heavier. A simple-cubic packed arrangement of spheres, by no means the most closely-packed geometry possible, has a packing fraction of 0.52, whereas fibrous filters have packing fractions of only a few percent. This very open structure, which is important in ensuring effective depth filtration with low airflow resistance, exists because a fibrous filter can be produced with relatively few points of fibre–fibre contact and much unsupported fibre, as the electron micrograph of a fibrous filter in Fig. 1.1 shows. The figure also shows that the structure is neither highly ordered nor completely random; and limited order

12

such as this is much more difficult to describe or to quantify than is either complete order or complete randomness. Filters can, however, be classified into structural types according to their method of manufacture, which leads us to consider paper filters, carded filters, porous foams and model filters.

Paper filters

Manufacture

Filters can be produced by modifying the normal paper-making process. Ordinary paper is made by allowing a slurry of fibres and irregular material to settle from suspension on to a porous base, which allows the fluid to drain away and the deposited material to dry out and form a sheet of paper. If the packing fraction of the paper is sufficiently low, it can function as a filter. The fibres, which may be cellulose or glass, will tend to lie in the plane of the paper but will not show a strong tendency towards any particular angular pattern. An electron micrograph of such a filter is shown in Fig. 2.1.

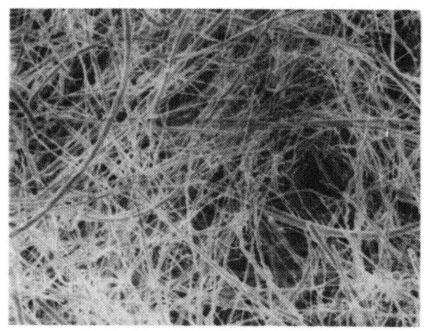

Fig. 2.1   Electron micrograph of a paper, air-laid filter.

The most efficient paper filters are made from glass fibre, the grade being varied by the use of fibres of different diameters. The fibres, which are dispersed in a slightly acid solution, are allowed to settle on to a moving porous belt. The surplus water is sucked out and the residue is dried on heated rollers to form a continuous sheet of filter paper.

It is clear that if the fibres of a paper filter are much longer than the filter is thick, they will tend to lie in or close to the plane of the paper. If the deposition process were completely random there would be no preferred direction in the plane, though external constraints may modify this. For instance if the paper suffers a slight stretching as it passes along the rollers, a correspondingly slight degree of fibre alignment will be imparted. Grain without fibre alignment can be produced if the porous belt on which the paper is made carries a pattern with a directional element, which would be transferred to the filter.

Alternatively paper filter materials can be produced by air laying (2). Glass microfibres can be spun by air blowing or flame attenuation, and then applied to a collector, along with binding agent, as part of a continuous process.

Description of structure

A fundamental study of the structure of filters (3) considers a thin slice, taken in a plane parallel to the filter surface, and regards this as typical in structure of the filter as a whole. An approximation to this structure is a series of straight widthless lines, random in position and orientation, drawn in the plane. The fibre sections in real filters will have a variety of lengths because the fibres themselves will have a variety of orientations, but an approximation may be made that the model fibres are all long.

A computer-generated picture of such a structure is shown in Fig. 2.2, and a section taken in one plane only through a real filter is shown in Fig. 2.3, illustrating the basic similarity (4). An analysis of the structure considers a number of questions: what is the probability of two fibres drawn at random intersecting within the finite area of the filter; how is the total number of pores formed related to the area density of fibres; and what are the distributions of pore size, pore shape and pore hydraulic radius.

The typical pore shape can be predicted by a simple argument. If the lines are widthless, the probability of more than two lines intersecting at a single point approaches zero. When two lines intersect, four angles are formed, each of which is the vertex angle of one of the polygons forming the pores of the filter. The average size of the four angles, and therefore of every angle, is 90°, and this means that the average number of sides of a pore is four, though of course triangular pores will be

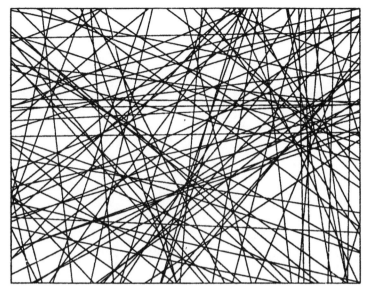

FIG. 2.2 Computer-generated model of paper filter, with random fibres.

relatively common, comprising in some filters 35% of the total number
(5), and pentagons and higher order polygons will also be frequent.

The probability of intersection, I, of two fibres of length, l, drawn at
random in a square element with side-length, l, is,

$$I = \frac{2}{\pi}. \tag{2.1}$$

If there are N_f fibres per unit area of filter the likely number of pores
produced by their intersections, N_p, is,

$$N_p = \frac{N_f(N_f - 1)}{\pi} - \frac{N_f}{2} \tag{2.2}$$

which, for large values of N_p, approaches,

$$N_p = \frac{N_f^2}{\pi}. \tag{2.3}$$

Predicting the size distribution of the pores is a more difficult
problem, and a simple analytical approximation is not possible. A
cursory examination shows that the distribution of the pore areas is
positively skewed and unimodal, and so it is reasonable to approximate
it by one of the standard distributions that have these properties: the

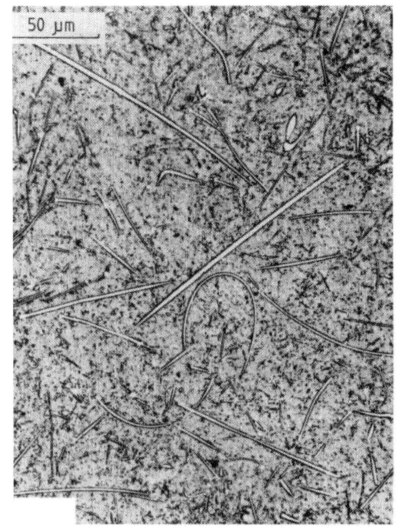

FIG. 2.3   Section through a paper filter in a plane parallel to its surfaces. (© Crown
Copyright; by permission of Her Majesty's Stationery Office.)

gamma distribution, the Weibull distribution and the lognormal
distribution.

Typical area distributions have the shape illustrated in Fig. 2.4, and
are clearly not lognormal (since cumulative lognormal distributions
appear as straight lines on graphs such as this). On the other hand, the
distribution of pore order (the number of sides to a polygonal pore)
does appear to be well approximated by a lognormal curve.

The process of dividing a filter up into pores of this sort is a random

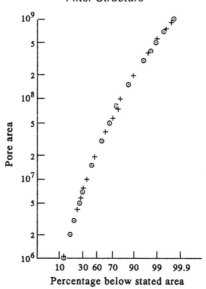

FIG. 2.4 Cumulative distribution of pore size: + Computer simulation; ⊙ fitted Weibull distribution. (© Crown Copyright; by permission of Her Majesty's Stationery Office.)

fragmentation process, similar to the comminution of solids that takes place during milling, or the formation of liquid emulsions. One-dimensional random fragmentation can be treated analytically (6), with the result that the distribution of lengths produced is exponential. Reference to fundamental statistical arguments (7) shows that if the likelihood of a multi-dimensional unit being bisected by an additional fragmentation is independent of its size, the resultant distribution should be lognormal. If the probability is proportional to the size, in this case the area of the pore, which is a more reasonable approximation, the result should be a Weibull distribution, given by,

$$W(d) = \frac{\delta_w}{\eta_w}\left(\frac{d}{\eta_w}\right)^{\delta_w-1} \exp\left[-\left(\frac{d}{\eta_w}\right)^{\delta_w}\right]. \tag{2.4}$$

A Weibull distribution, fitted to data, is shown in Fig. 2.4 (8), which shows that theory and numerical computation agree.

The influence of a pore on airflow is governed more by its hydraulic radius than its geometric area. The hydraulic radius is twice the quotient of the cross-sectional area and the circumference, and so it equals the geometric radius for a circular pore but it is rather smaller for pores of any other shape. The flow of air through long pores depends critically on this parameter, though the significance in the system under examination is likely to be smaller.

Fractal dimension

The computer-modelled structure consists of pores of a range of sizes, which, taken together, completely fill the area of the filter. If, however, only pores larger than a chosen area, the area of resolution, are considered, there will be a residual area, which will decrease with the area of resolution. This feature is shared with classic fractal figures like the Sierpinski carpet (9). The fractal dimension of such figures is given by the relationship between the two areas, and in the classic fractal structure the dimension does not vary with the area of resolution. In the random filter the fractal dimension, which cannot exceed 2, can be measured from a computer simulation (10), or calculated from the Weibull distribution in equation 2.4, and in both cases it decreases with the area of resolution. In the latter case limiting values for high and low areas of resolution are 2 and $2-2\delta_w$ respectively.

Carded filters

Manufacture

Carding is a common textile process in which fibres are repeatedly combed with metal hooks, whereby knots and clumps of fibres are disentangled. In most textile manufacture the carded fibres are spun into a yarn or made into a compact felt; but a moderate compaction of the mass of carded fibres produces a structure suitable for a filter. The stress to which the fibres are subjected during carding is quite high, and fibres smaller than about 15 μm in diameter are generally too weak to withstand it. The mass of fibres resulting from carding consists of fibres that are roughly aligned, which shows itself by the material being easier to pull apart in one direction than in the direction perpendicular to it, behaviour shown by cotton wool, which is carded viscose.

Natural fibres like wool or cotton are not straight, but crimped. Synthetic fibres are straight when they are produced, but crimp is usually applied to them because carding straight fibres tends to produce a dense, irregular and weak structure that is not particularly suitable either for filters or for textiles. The crimp of some of the fibres and the general direction of fibre alignment are apparent in the carded filter shown if Fig. 1.1. However, the structure illustrated will not be typical, since Fig. 1.1 shows a cut surface, and some degree of structural relaxation is inevitable in this situation.

Description of structure

The microscopic structure of compacted filters can be examined if they are potted and sectioned (11), but a problem in the study of filters with a low packing fraction is to do this without altering their structure. A method of fixing them with glue before potting, followed by polishing, has been successfully used, and a result, in the form of a section perpendicular to the direction of fibre alignment, is shown in Fig. 2.5 (12).

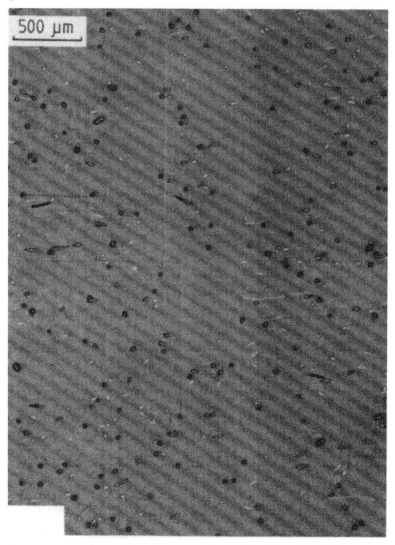

FIG. 2.5   Section perpendicular to the direction of fibre alignment in a carded filter.
(© Crown Copyright; by permission of Her Majesty's Stationery Office.)

The structure shown in the figure looks approximately random but it does not have all of the required properties of a truly random structure. In a random structure the number of fibres counted in a series of fields of equal size should follow a Poisson distribution, with the mean number of fibres equal to the area density and the variance equal to the mean (13). If this process is applied to the structure shown in Fig. 2.6a, which is a clumped structure, the variance will be greater than the mean, provided that the fields are large enough to contain several fibres. On the other hand if the structure in Fig. 2.6b, an ordered structure, is studied, the variance will be less than the mean. Applying the method to observed structures like that in Fig. 2.5 gives rather larger variances than means, indicating that the structures are not quite random but slightly clumped.

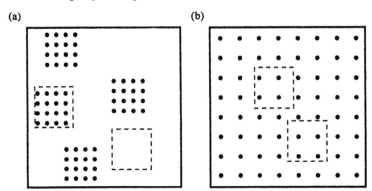

FIG. 2.6 Idealised filter structures: (a) idealised clumped filter; (b) idealised ordered filter. (© Crown Copyright; by permission of Her Majesty's Stationery Office.)

An alternative method is to measure the nearest neighbour distances and to plot the result as a cumulative distribution. The result for a random distribution can be calculated analytically (14). The probability $P_N(r)$ that the nearest neighbour distance is less than r, is,

$$P_N(r) = 1 - \exp(-\mu_F \pi r^2) \qquad (2.5)$$

where μ_F is the mean number of sectioned fibres per unit area. The clumped distribution will show itself by having rather smaller nearest neighbour distances, and the ordered rather larger, and, as expected, the distribution in Fig. 2.5 has smaller nearest neighbour distances than does the random distribution.

Filters with short fibres

Certain carded filters are made from fibres that are rather shorter than ideal, and the result of this is an irregular structure containing small regions of high fibre density called "neps". A section through such a filter is shown in Fig. 2.7, in which a nep is clearly visible towards the bottom centre of the figure.

Felting

Carded filters generally lack mechanical strength, being about as strong as cotton wool. This is no problem if the filters are used in capsules or other containers, but it makes them unsuitable for use in sheet form. A common way of improving the strength of the material itself is by felting (the words "felt" and "filter" have the same origin). The surfaces of wool fibres have scales, which hold the fibres together when the material is felted; and wool can be felted by the application of heat, humidity and pressure, a process dating back to antiquity. Synthetic fibres, with their smooth surfaces, cannot be effectively treated in this way, but they can be felted by "needling", a process in which the carded material is punched by barbed needles, with a punch density typically of several tens per square centimetre. The barbed needles push some of the fibres through the material causing tangling, generally resulting in improved strength but often with diminished filtration efficiency. The surface of material treated in this way frequently shows patterns of small dimples where the needles have entered. A section through a needlefelt is shown in Fig. 2.8.

Porous foam

Porous foams are non-fibrous depth filters. The filter matrix of a porous foam, unlike that of a fibrous filter, is continuous, which makes it easier to remove dust from the filter for the purpose of cleaning or analysis. The rough treatment incurred during the cleaning of a carded fibrous filter or a paper filter usually results in fibre loss and deterioration of structure.

Figure 2.9 is an electron micrograph showing that the foam is made up of a large number of tightly-packed bubbles, which have been made to burst at points of contact, producing a single ramifying and anastomosing space within a matrix formed from the residual material. The geometric arrangement of the bubbles is not clear from Fig. 2.9, but a section of the foam, shown in Fig. 2.10, gives a clue to the structure. Contiguous bubbles, in threes, are typically separated by triangular regions of the constituent material. The typical number of these triangles bordering on any section of a bubble is about six; and this indicates that the structure is relatively close-packed, probably resembling face-centred cubic. A loosely packed structure would have fewer sections of solid border to each bubble; for example a simple-

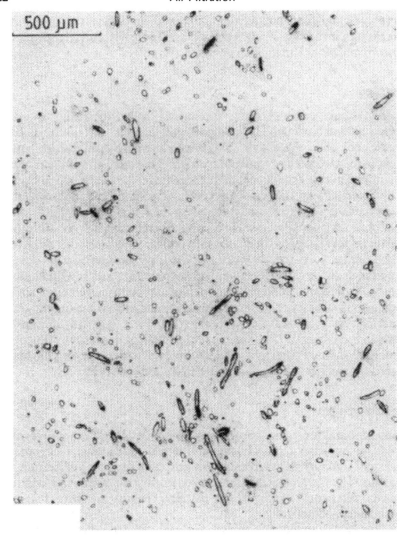

Fig. 2.7 Section through a carded filter containing short fibres, showing a "nep". (© Crown Copyright; by permission of Her Majesty's Stationery Office.)

cubic packed structure would be unlikely to have more than four. In addition, during the formation of the foam, the tendency of the material to adopt a configuration with a low surface area and, therefore, a low total surface energy, would favour a relatively close-packed structure.

A special feature of this type of material is that it is produced in a range of grades, which vary only little in packing fraction or in geometry, but by more than an order of magnitude in average bubble size. This not only has practical value but also provides a useful system

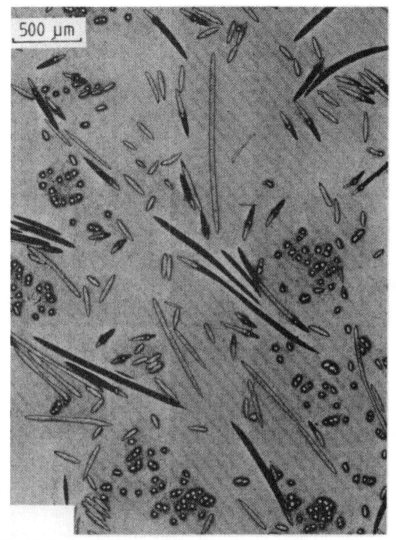

FIG. 2.8 Section through a needled filter, illustrating clumps of fibres at right angles to the section, where needles have penetrated. (© Crown Copyright; by permission of Her Majesty's Stationery Office.)

by which the validity of certain scale-dependent theoretical assumptions can be examined.

Model filters

Both experiments on filters and practical implementation of them involve filters with the irregular structure illustrated so far. However,

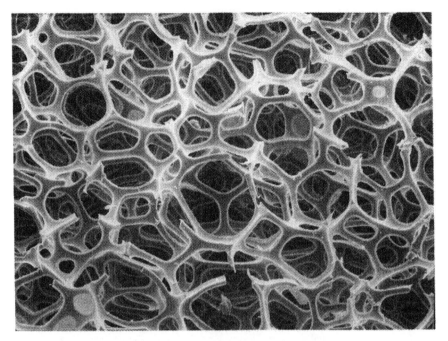

FIG. 2.9 Electron micrograph of a porous foam filter.

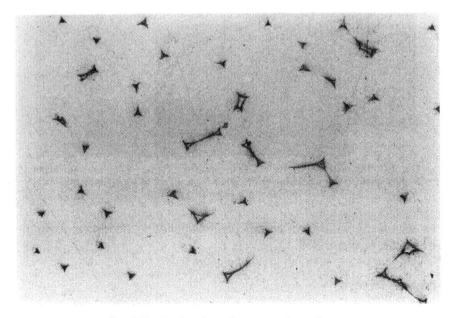

FIG. 2.10 Section through a porous foam filter.

filtration theory requires approximations to filter structure that are sufficiently simple to form the basis of calculation but sufficiently complicated to give a realistic description of filters used in practice. It will be seen below that the most widely used theoretical models of fibrous filters are regular periodic arrays, in particular two-dimensional fibre arrays in which conditions are assumed not to vary along the length of a fibre. This approximation is imperfect but likely to be better for carded filters, in which the fibres are relatively aligned, than for paper filters. The approximation must break down at points of fibre–fibre contact, though these are relatively infrequent.

Experimental model filters can be produced with structures of very high symmetry, so that their observed behaviour can be approximated very closely, or even described exactly, by analytical theories. Results from experiments on model filters therefore form a valuable bridge between theory and practice. A number of experimental model filters have been used in studies of filtration.

Standard sieves

The material of fine standard sieves, which are readily available, is produced to a high level of precision, and it forms a ready made single layer filter. By putting together a number of sieves, using properly constructed spacers, a well-defined filter can be produced (15, 16). Figure 2.11 shows an example of such sieve material.

A rather similar alternative to the use of metal sieves is that of woven polymer-fibre fabrics, which have inter-fibre spacings of the order of micrometres.

Wound wire

The structure of sieves is simple but not two-dimensional, and so the connection between theory and experimental results with them is not quite rigorous. Model filters without this problem can be made from wire, which can be produced in a variety of thicknesses down to a few micrometres. It is possible, if care is taken, to wind wire around a frame on a lathe with a prescribed spacing between individual wires (17, 18). The frame will then have a double layer of wire, but the wire may be glued to it and the two layers separated to give the model filter element. Several layers may be put together, producing a two-dimensional fibre array. The disadvantage of making model filters in this way is that it is highly labour-intensive.

A filter can be assembled from a number of layers produced in this way, so that although the wires in each individual layer are parallel, those in adjacent layers are orientated at random, giving rise to the so-

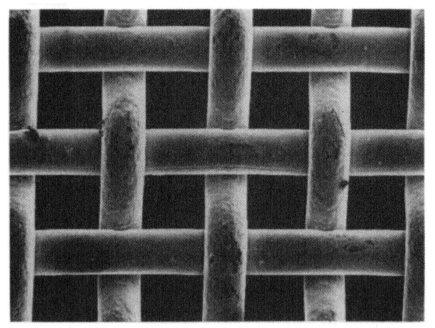

FIG. 2.11   Electron micrograph of fine sieve material.

called "fan-model" filter (19), which has formed a useful link between model filters of parallel fibres and real filters.

Photo-etched filters

A method that is much easier to use than the above is that of the photo-etching of elements of a filter from a thin metal sheet. Once a master design with the required structure has been produced, this can be easily reproduced to give a multi-layer filter (20). The limitations are that it is difficult to produce really fine fibres, less than about 500 μm in diameter, by this technique, and that the fibres produced are not circular in cross-section, but rather closer to hexagonal, as illustrated in Fig. 2.12.

Lithographically produced filters

Finer filters can be produced from silicon by a lithographic technique (21) similar to the methods used to produce micro-chips. The filters produced have the advantage of those described above in that mass-production of layers of the filter is possible, but the methods described are capable of producing finer fibres. Two techniques, of electroplating and of diffusion-etching, produce fibres of rectangular and of elliptical cross-section respectively.

FIG. 2.12   Electron micrograph of a photoetched model filter.

References

1. KITTEL, C. *Introduction to Solid State Physics*, Wiley, New York, 1953.
2. KANAGAWA, A. An engineered glass microfiber media for filtration applications. *Advances in Filtration and Separation Technology* (ed. K. L. RUBOW), pp. 341–345. American Filtration Society, 1991.
3. PIEKAAR, H. W. and CLARENBURG, L. A. Aerosol filters — pore size distribution in fibrous filters. *Chem. Eng. Sci.*, 1967, **22**, 1399–1408.
4. WILKINSON, E. T. and DAVIES, G. A. A stochastic model for the filtration of dilute suspensions using non-woven cloths. *Can. J. Chem. Eng.*, 1985, **63**, 891–902.
5. ABDEL GHANI, M. S. and DAVIES, G. A. Simulation of non-woven fibre mats and their application to coalescers. *Chem. Eng. Sci.*, 1985, **40**(1), 117–129.
6. TENCHOV, B. V. and YANEV, T. K. Weibull distribution of particle sizes obtained by uniform random fragmentation. *J. Coll. Int. Sci.*, 1986, **111**(1), 1–7.
7. KOLMOGOROV, A. N. The logarithmically normal distribution of dimensions of particles when broken into small parts. *Dokl. Akad. Nauk SSSR*, 1941, **31**(2), 99–101 (in Russian). English translation N69-29262 (NASA-TT-F-12287).
8. BROWN, R. C. The pore size distribution of model filters produced by random fragmentation described in terms of a Weibull distribution. *Chem. Eng. Sci.*, in the press.
9. CRILLY, A. J., EARNSHAW, R. A. and JONES, H. *Fractals and Chaos*, Springer-Verlag, New York, 1991.
10. KAYE, B. H. Describing filtration dynamics from the perspective of fractal geometry. *KONA Powder and Particle*, 1991, **9**, 218–236.
11. SCHMIDT, E. and LOEFFLER, F. Preparation von Staubkuchen. *Staub Reinhalt. Luft.*, 1989, **49**, 429–432.
12. VAUGHAN, N. P. and BROWN, R. C. Measurement of filter structure. *J. Aerosol Sci.*, 1992, **23**, S741–S744.
13. ROLLIN, A. L., DENNIS, R., ESTAQUE, L. and MASOUNAVE, J. Hydraulic behaviour of synthetic non-woven filter fabrics. *Can. J. Chem. Eng.*, 1982, **60**(2), 226–234.

14. WAKER, A. J. and BROWN, R. C. Application of cavity theory to the discharge of electrostatic dust filters by X-rays. *Appl. Radiat. Isot.*, 1988, **39**(7), 677–684, *Int. J. Radiat. Appl. Instrum. Part A.*

15. WALKENHORST, W. On some properties of a filter with high intake velocity. *Staub Reinhalt. Luft.*, 1973, **33**(4), 162–167 (in English).

16. EMI, H., WANG, C. S. and TIEN, C. Transient behaviour of aerosol filtration in model filters. *AIChE J.*, 1982, **28**(3), 397–404.

17. TSIANG, R. C., WANG, C. S. and TIEN, C. Dynamics of particle deposition in model filters. *Chem. Eng. Sci.*, 1982, **17**(11), 1661–1673.

18. KANAOKA, C. and HIRAGI, S. Pressure drop of air filter with dust load. *J. Aerosol Sci.*, 1990, **21**(1), 127–137.

19. KIRSCH, A. A. and FUCHS, N. A. Studies in fibrous aerosol filters II. Pressure drops in systems of parallel fibres. *Ann. Occ. Hyg.*, 1967, **10**, 23–36.

20. BROWN, R. C. A many-fibre theory of airflow through a fibrous filter II: Fluid inertia and fibre proximity. *J. Aerosol Sci.*, 1986, **17**, 685–697.

21. BERGMANN, W. and CIARLO, D. Fabricating micro air filters using lithographic techniques. *International Symposium/Workshop on Particle and Multiphase Proceedings, and 16th Annual Meeting of the Fine Particle Society*, 1985, *April 22–26, Miami, Florida.*

CHAPTER 3

Flow Patterns and Pressure Drop

Nature of airflow through a filter

The next point to consider after filter structure is the pattern adopted by the flowing air. This will depend on both the structure of the filter and the nature of the flow. Once the basic pattern is known, the pressure drop across the filter can be calculated, and when it is known in detail the behaviour of entrained particles can be derived, leading to a better understanding of particle capture mechanisms and enabling values of single fibre efficiency to be calculated.

Some understanding of both flow pattern and pressure drop follows from simple arguments. The behaviour of the air flowing past the fibres of a filter can be affected by four of its intrinsic properties: its mass or inertia; its viscosity; its elasticity; and its molecular properties.

The pressure drop across a filter is often only a few millimetres of water, a very small fraction of atmospheric pressure. This means that it is reasonable to assume that the air is incompressible; and this approximation will be assumed to hold throughout.

Air can be treated as a continuous fluid provided that the obstacles in its path are large compared with the mean free path of air molecules, which is 0.065 μm at NTP. This means that molecular effects in airflow can be neglected provided that the fibres are not of submicrometre size. For the time being this will be assumed, though a correction for molecular effects will follow later in the chapter. However, aerosols commonly contain submicrometre particles and so it may be necessary to include molecular effects in a study of particle dynamics applied to filtration, even in situations where the effects can be neglected during a study of the airflow.

Whether the inertia of the air or its viscosity dominates when air flows past an obstacle depends on the scale of the system and on the velocity of the air. The relative importance of the two effects depends on the size of a dimensionless parameter, the Reynolds number, $Re(1)$,

$$Re = \frac{2R\rho_g U}{\eta} \tag{3.1}$$

where ρ_g, the density of the air, is approximately 1.20 kg m^{-3}, and η, its coefficient of viscosity is, $1.81 \times 10^{-5} \text{ kg m}^{-1} \text{ s}^{-1}$ at NTP. R is the radius of a fibre, or $2R$ can be considered equivalent to the width of any obstacle, and U is the air velocity.

According to equation 3.1, the Reynolds number for air passing a human body at a velocity of 2 ms^{-1}, a gentle breeze, will be about 60,000. That for air passing a textile fibre with a diameter of $20 \ \mu\text{m}$ at a velocity of 0.10 ms^{-1}, common conditions in a filter, is 0.14. The difference in Reynolds number illustrates that the nature of the flow is very different in the two cases; the former is dominated by the inertia of the air, the latter by its viscosity.

Steady state flow and Stokes flow

In most instances airflow through a filter can be well-approximated by Stokes flow, in which inertia is neglected completely (the Reynolds number is assumed to be zero). For the time being we shall make this approximation, though a correction for low but finite inertia of the air will be made later.

Stokes flow is one example of steady state flow; the flow pattern does not vary with time. Flow perpendicular to the axis of a fibre is two-dimensional flow, which means that it does not vary along the fibre axis but only in the two dimensions perpendicular to the axis; all sections through the fibre perpendicular to its axis will look the same.

A particularly simple means of describing two-dimensional steady-state flow uses the stream function (2), ψ. The value of ψ, which is a function of position only, at any point is a measure of the air flowing between that point and some arbitrary origin. There will be no airflow across lines joining points at which ψ is constant; such lines are streamlines, as shown in Fig. 3.1. Streamlines illustrate the flow pattern and they show both the direction of the flow and its speed. Closely spaced streamlines indicate rapid flow; widely spaced streamlines indicate slow flow. Moreover, ψ is simply related to the velocity of the air, by means of the following equations, in cartesian co-ordinates,

$$U_x = \frac{\partial \psi}{\partial y}; \quad U_y = -\frac{\partial \psi}{\partial x} \tag{3.2}$$

and in polar co-ordinates,

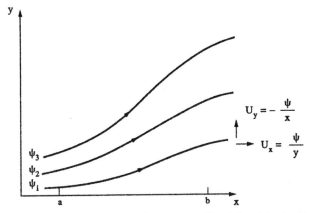

FIG. 3.1   General illustration of streamline pattern. The airflow is in the direction of the arrows, and so $\psi_1 < \psi_2 < \psi_3$. U_x is greater than $x = a$ than at $x = b$ because at a the streamlines are more closely spaced.

$$U_r = \frac{1}{r} \frac{\partial \psi}{\partial \theta}; \quad U_\theta = -\frac{\partial \psi}{\partial r}. \tag{3.3}$$

The simplest example of the form taken by ψ is the stream function describing uniform flow, at velocity U, in the x-direction. This follows from equation 3.2,

$$\psi = Uy. \tag{3.4}$$

For any flow pattern that can be described by streamlines, the air velocity obeys the equation of continuity.

$$\nabla \cdot \underline{U} = \frac{\partial U_x}{\partial x} + \frac{dU_y}{\partial y} = \frac{\partial U_r}{\partial r} + \frac{U_r}{r} + \frac{1}{r} \frac{\partial U_\theta}{\partial \theta} = 0. \tag{3.5}$$

Vector fields that obey equation 3.5 are termed solenoidal. The flow field for incompressible airflow is solenoidal, and so are the field of gravity and any electric field. The special significance of solenoidal fields in calculations of the capture of particles will become apparent in Chapters 4 and 6.

A further important property of Stokes flow is that the flow pattern does not alter with the velocity. For example, the parabolic velocity profile in Poiseuille flow between planes or through a capillary alters only in scale as the velocity is changed. Altering the macroscopic velocity causes the microscopic velocity at every point to be scaled by a common factor. If the scaling factor is negative, the direction of macroscopic flow is reversed, and this means that in Stokes flow around any object with upstream–downstream symmetry, like a cylindrical fibre perpendicular to the flow direction, the flow pattern will have the same upstream–downstream symmetry. The symmetry and velocity invariance are distinguishing features of Stokes flow.

Simple model of airflow around a filter fibre

The concepts developed above can be used to deduce the flow pattern around a filter fibre. A section through the fibre is shown in Fig. 3.2. Polar co-ordinates are used for the description of a macroscopic flow, which is assumed to be from left to right. The air in contact with the fibre is stationary, which is the normal Stokes flow condition, but at a small distance, $\delta = r - R$ from the fibre, the tangential flow will be as the arrows indicate.

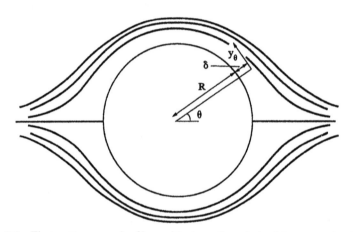

Fig. 3.2   Flow pattern round a fibre, with streamlines derived from equation 3.9.

The figure has reflection symmetry across the diameter in the direction of macroscopic flow, and so U_θ must have the same symmetry.

$$U_\theta(-\theta) = -U_\theta(\theta) \qquad (3.6)$$

Furthermore, Stokes flow has upstream–downstream symmetry and so,

$$U_\theta(\pi - \theta) = U_\theta(\theta). \qquad (3.7)$$

The simplest function that satisfies these symmetry requirements is $\sin\theta$ (and the general function is $\sin(2n-1)\theta$). The velocity must vanish at $\delta = 0$, and so the simplest form it can take is,

$$U_\theta = -2\Upsilon(r-R)\sin\theta. \qquad (3.8)$$

Applying equation 3.8 to equation 3.3 and assuming for convenience that $\psi = 0$ at the fibre surface, we have for the stream function,

$$\psi = \Upsilon(r-R)^2 \sin\theta \qquad (3.9)$$

and for the radial velocity,

$$U_r = \frac{\Upsilon}{R}(r-R)^2 \cos\theta. \qquad (3.10)$$

Streamlines corresponding to equation 3.9 are plotted in Fig. 3.2. These, like the expression in the equation, are correct only close to the fibre surface, but they do indicate the important features of the flow pattern, in particular that the streamlines diverge as they approach the fibre, and that the region of space surrounding the fibre is a region of relative stagnation.

Simple description of pressure drop

Some understanding of pressure drop also follows from the Stokes flow conditions, but in this case dimensional analysis (3) must be used. It can reasonably be assumed that the pressure drop across a filter varies with its thickness, h, the macroscopic air velocity, the fibre radius, R, and the only significant intrinsic parameter, the coefficient of viscosity of the air. The packing fraction, c, is clearly important, but this can be formally included by means of an unknown dimensionless function, f, and the rest of the analysis can then proceed, with the assumption that the structure is constant. Moreover, it is almost axiomatic that the pressure drop across a filter is directly proportional to its thickness, which means that,

$$\Delta p = hU^\alpha \eta^\beta R^\gamma f(c). \qquad (3.11)$$

Equating the dimensions of the two sides enables simple equations in the powers to be obtained, and the result is,

$$\Delta p = \frac{\eta h U f(c)}{R^2}. \qquad (3.12)$$

Equation 3.12 embodies the first law of filtration theory, Darcy's

Law (4), which states that the pressure drop across a filter is proportional to the rate of flow of fluid through it.

General fluid dynamics theory

The elementary approach adopted above must now be left behind, and those aspects of fluid dynamics that will be used later in the chapter must be considered. The treatment is simplified and condensed, and any reader requiring mathematical rigour is referred to the source text books (2, 5, 6). Part of the section deals with flow in which inertia is finite, because this is needed for the treatment of theories of flow with an inertial component. Other parts apply only to Stokes flow, and these are generally more important, for most of the theoretical models of filters deal with Stokes flow.

Elementary theory of viscosity (1) states that the shearing stress at any point—where the velocity gradient perpendicular to the direction of fluid motion is dU/dz—is directly proportional to the value of the gradient, so that the frictional force per unit area is given by,

$$F_s = \eta \, \frac{dU}{dz}. \tag{3.13}$$

The force acting can be related to the pressure gradient, and so equation 3.13 can be generalised to give, in vector notation,

$$\nabla p = \eta \nabla^2 \underline{U}. \tag{3.14}$$

Equation 3.14 applies only to the situation where viscosity is dominant. In a complete theory of fluid flow the motion of a continuous fluid is considered to obey the Navier–Stokes equation (3); but for the moderately complicated situation of steady-state incompressible flow, the Navier–Stokes equation can be written in a simplified form,

$$(\underline{U} \cdot \nabla)\underline{U} = \frac{\eta}{\rho_g} \nabla^2 \underline{U} - \frac{1}{\rho_g} \nabla p. \tag{3.15}$$

When inertia can be neglected equation 3.15 reverts to the simpler form, equation 3.14.

If, now, we take the curl of each side of equation 3.15, the gradient terms vanish. Substituting for ψ according to equation 3.2 gives the equation describing the stream function in steady-state two-dimensional incompressible flow,

$$\frac{\eta}{\rho_g} \nabla^4 \psi = \frac{\partial \psi}{\partial y} \frac{\partial \nabla^2 \psi}{\partial x} - \frac{\partial \psi}{\partial x} \frac{\partial \nabla^2 \psi}{\partial y}. \tag{3.16}$$

Two functions, describing the flow of a fluid, that are particularly useful in filtration theory are the vorticity and the stress tensor. The vorticity, ω, is the curl of the velocity, and is a measure of the tendency of fluid to rotate or to cause rotation of suspended bodies during flow. In the special case of two dimensional flow it takes the following form, in polar coordinates,

$$\omega = \frac{\partial U_\theta}{\partial r} - \frac{1}{r} \frac{\partial U_r}{\partial \theta} + \frac{U_\theta}{r} \tag{3.17}$$

in terms of ψ, ω is given by,

$$\omega = -\nabla^2 \psi. \tag{3.18}$$

Substitution of equations 3.2 and 3.18 into equation 3.16, gives the diffusion equation which describes the behaviour of the vorticity.

The general component of the second function, the stress tensor, is σ_{ij}, which describes the stress, in direction j, acting on a unit area of surface whose normal is in direction i. In polar co-ordinates the tangential stress, $\sigma_{r\theta}$, is given by,

$$\sigma_{r\theta} = \eta \left(\frac{1}{r} \frac{\partial U_r}{\partial \theta} + \frac{\partial U_\theta}{\partial r} - \frac{U_\theta}{r} \right) \tag{3.19}$$

and the normal stress, σ_{rr}, by,

$$\sigma_{rr} = -p + 2\eta \frac{\partial U_r}{\partial r}. \tag{3.20}$$

Microscopic airflow patterns in filters

The earliest theories of microscopic viscous flow through fibrous filters sought descriptions in terms of theories existing at the time: the flow of air around an isolated fibre and flow through granular media. Neither of these approaches met with complete success. Isolated fibre theory starts with a simplification of equation 3.16. The stream function is then assumed to comprise two components, one relating to uniform flow, and given by equation 3.4, and the other a perturbation caused by the fibre. Simplifying the results gives the equation of Oseen (7):

$$\left(\nabla^2 - \frac{U\rho_g}{\eta}\frac{\partial}{\partial x}\right)\nabla^2\psi = 0. \tag{3.21}$$

A number of workers (8, 9, 10) have solved the equation with varying levels of detail. All workers agree that the drag per unit length of fibre caused by the flow is,

$$F_d = \frac{4\pi\eta U}{(2.0022 - \ln(Re))}. \tag{3.22}$$

The drag acting on a unit length of fibre is related to the pressure drop across a filter of thickness h containing a length L of fibre per unit volume by,

$$\Delta p = F_d L h = \frac{chF_d}{\pi R^2}. \tag{3.23}$$

Flow patterns are always much more difficult to calculate than drag or pressure drop, but an approximation to the flow pattern has been obtained on the basis of this model (10). A feature of isolated fibre flow patterns is that a weak vortex exists on the downstream side of the fibre, as illustrated in Fig. 3.3, and that the size of the vortex increases

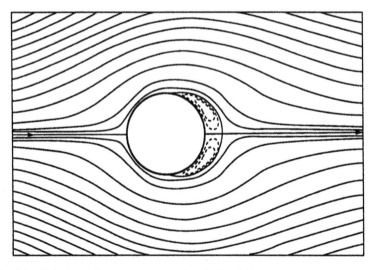

FIG. 3.3   Calculated flow pattern around an isolated fibre (10). A vortex exists, on the down-stream side making the flow asymmetric. (With permission of Oxford University Press.)

with flow velocity. Expressed more generally, these features are that the flow pattern has velocity-dependence and that it lacks upstream–downstream symmetry.

Isolated fibre theory has been applied in the interpretation of certain experiments carried out on fibres removed from filters and studied in isolation (11), but it is not a suitable description of fibres within filters, for the pressure drop predicted by it does not have the simple velocity-dependence of Darcy's law. Typical interfibre distances in a filter are only tens of micrometres, and neglecting the effects of nearby fibres is not a valid approximation.

The theory used to describe flow through granular beds (12, 13) does not have this drawback, but rather goes to the opposite extreme, treating only dense porous media. A detailed account of granular filtration is available (14), but it is sufficient for our purpose that the theory considers filter material to be composed of a series of ramifying and anastomosing pores, and that the treatment is a logical progression from the Poiseuille theory of flow through capillaries (1). A section through granular material reveals structures that are pores or incomplete pores, in which a cavity is surrounded by the surfaces of several touching grains, but the photographs of sections through fibrous filters, illustrated in the previous chapter, show that the structure of these is fundamentally different.

Nearly all successful theories of filtration through fibrous filters consider at the outset that the filter is neither densely packed nor composed of isolated fibres, and that the packing fraction has an intermediate value. In addition it is almost always assumed that the flow through the filter is dominated by viscosity (that the Reynolds number is very small). If the approximation is made that *Re* is zero, equations 3.16 or 3.21 can be reduced to a simple form, the biharmonic equation:

$$\nabla^4 \psi = 0. \tag{3.24}$$

Filters approximated by single layers of fibres

Although no solution exists for Stokes flow round an isolated fibre, a solution does exist for the case of flow through a sheet made up of equally spaced fibres parallel to each other and perpendicular to the airflow. The solution of the problem relies on the use of complex variable theory (2, 15, 16, 17). The complex velocity, given by

$$W = U_x - iU_y \tag{3.25}$$

satisfies the Stokes flow equation, and can be expressed as a series,

$$\frac{W}{U} = 1 + a_o \left\{ \ln(2 \sinh \bar{w}) + \ln(2 \sinh w) - (w + \bar{w}) \coth w \right.$$

$$+ \sum_{n=1}^{\infty} a_{2n} \left[\frac{d^{2n-1}}{dw^{2n-1}} \coth \bar{w} - (w + \bar{w}) \frac{d^{2n}}{dw^{2n}} \coth w \right]$$

$$\left. + \sum_{n=1}^{\infty} b_{2n} \frac{d^{2n-1}}{dw^{2n-1}} \coth w \right\} \tag{3.26}$$

where,

$$w = \frac{\pi}{2b}(x + iy). \tag{3.27}$$

$2b$ is the interfibre spacing as shown in Fig. 3.4. A description of the flow pattern requires the calculation of a sufficient number of coefficients for the residue to be negligible, but calculation of the drag per unit length of a fibre requires only a_o, and if an expansion of a_o is carried out, the drag per unit length of fibre can be expressed as,

$$F_d = \frac{4\pi\eta U}{\left[-\ln\left(\frac{R}{2b}\right) - 1.34 + \frac{\pi^2}{3}\left(\frac{R}{2b}\right)^2 + \cdots \right]}. \tag{3.28}$$

The values of the coefficients are given, up to the 14th order (16), and the flow pattern illustrated in Fig. 3.4 is calculated from these.

If a filter is considered to be a stack of such sheets, as shown in Fig. 3.5, and if it is assumed that each sheet of fibres contributes to the pressure drop independently of the others, the pressure drop would be,

$$\Delta p = \frac{4ch\eta U}{R^2 \left[-\frac{1}{2}\ln(c) - 0.76 + \frac{\pi c}{3} - \frac{\pi^2 c^2}{18} + \cdots \right]}. \tag{3.29}$$

The validity of this expression has been put to the test (17), and it has been shown to give a good description of the pressure drop, provided that the separation of sheets is at least equal to the interfibre spacing

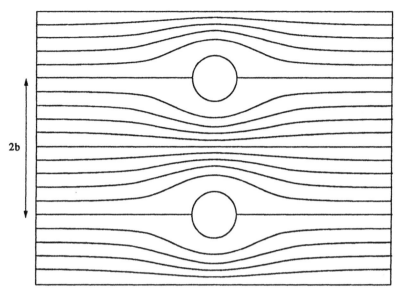

FIG. 3.4 Flow pattern through a layer of parallel fibres calculated according to complex variable theory (16). The flow pattern shows the upstream–downstream symmetry of Stokes flow.

within the sheets. Calculated values of $f(c)$, which are related to the pressure drop by way of equation 3.12, for a range of values of packing fraction, are given in Table 3.1. This model does not distinguish between the two structures in Fig. 3.5, which are called channel and staggered, but measured values of pressure drop on model filters of the two structures are similar unless the filters are compressed.

Single fibre theory

Single fibre theory must be distinguished from isolated fibre theory, for the latter considers a fibre in an effectively infinite space and it requires fluid inertia to be included in the solution from the start. The former, although focusing attention on one fibre, considers the effects of other fibres at the outset, in such a way that the Stokes flow equation, 3.24, can be solved, in cylindrical polar coordinates.

$$\left(\frac{\partial^2}{\partial r^2} + \frac{1}{r}\frac{\partial}{\partial r} + \frac{1}{r^2}\frac{\partial^2}{\partial \theta^2}\right)^2 \psi = 0 \tag{3.30}$$

Differential equations of this sort have an infinity of solutions, but

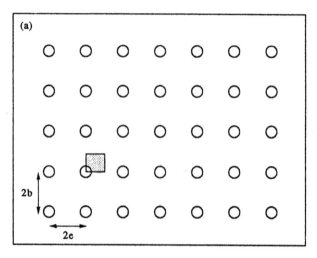

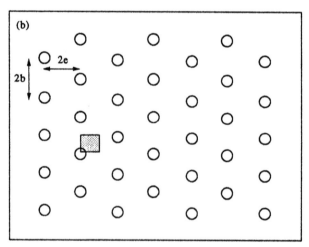

FIG. 3.5 Sections through arrays of cylindrical fibres illustrating: (a) channel model, and (b) staggered model. In each case the irreducible symmetry element is shown.

the simplest, which is the one that we need, is that for which ψ is proportional to $\sin \theta$,

$$\psi = g(r)\sin \theta = \left(Ar + \frac{B}{r} + Cr \ln\left(\frac{r}{R}\right) + Dr^3\right)\sin \theta. \qquad (3.31)$$

Cell models

Finding the unknown constants in equation 3.31 requires boundary

TABLE 3.1
Calculated values of f(c) *according to various flow models*

c	Miyagi	Happel	Kuwabara	Hasimoto	Variational Channel	Variational Stagger	Davies
0.001	0.00149	0.00135	0.00148	0.00147			0.00051
0.002	0.00341	0.00307	0.00339	0.00337	0.00344	0.00352	0.00143
0.005	0.01060	0.00931	0.0105	0.0104	0.0106	0.0109	0.00566
0.01	0.0259	0.0222	0.0256	0.0254	0.0256	0.0265	0.0160
0.02	0.0666	0.0549	0.0653	0.0646	0.0653	0.0682	0.0453
0.05	0.264	0.200	0.251	0.247	0.250	0.266	0.179
0.1	0.919	0.605	0.802	0.779	0.800	0.873	0.506
0.2	6.02	2.33	3.27	3.00	3.35	3.76	1.43

conditions to be applied. Two of these are quite straightforward, being that both the radial and the tangential components of the air velocity vanish at the fibre surface. The others require the model to be rather more specific, and this is normally achieved by defining a reference surface concentric with the fibre surface, as shown in Fig. 3.6, at such a

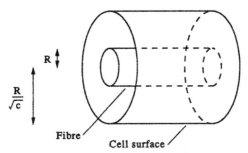

FIG. 3.6 Model used in cell theory. The radius of the cell enclosing the fibre is fixed by the packing fraction of the filter. In the example illustrated the packing fraction is 0.11.

distance that the packing fraction of the fibre within this cylinder is identical to that of the fibres within the filter as a whole; in other words that its radius R' is equal to the fibre radius divided by the square root of the packing fraction, c. Once this has been defined, it is possible to apply the third boundary condition, which is obtained by relating the value of the stream function at the $\theta = \pi/2$ point on the boundary to the flow velocity within the filter. The final boundary condition is not so obvious, and it is here that the two principal single-fibre theories, that of Happel (18) and that of Kuwabara (19), differ in their choice. The former requires the tangential stress as given by equation 3.19 to vanish, and the latter requires the vorticity as given by equation 3.18 to vanish, though other conditions could be imagined.

Because the flow equation is solved in a limited region, assumed to be typical of the filter as a whole, the Kuwabara and the Happel models are often called "cell models". The expressions for $g(r)$ given by the two models depend on the constants in equation 3.31, which are listed in Table 3.2. The pressure drop in each case can be derived directly from equation 3.31 using the theory of Imai (20), which is summarised by Pich (21). The value for Happel's theory is,

$$\Delta p = \frac{4ch\eta U}{R^2\left[-\frac{1}{2}\ln(c) - \frac{1}{2}\frac{(1-c^2)}{(1+c^2)}\right]} \tag{3.32}$$

and for Kuwabara's,

$$\Delta p = \frac{4ch\eta U}{R^2\left[-\frac{1}{2}\ln(c) - 0.75 + c - \frac{c^2}{4}\right]}. \tag{3.33}$$

TABLE 3.2
Constants in equation 3.31 according to cell theory

Constant	Happel	Kuwabara	Kuwabara at finite Kn
A	$\frac{-1+c^2}{2(1+c^2)}J$	$\frac{-1+c}{2}J$	$\frac{-1+c}{2}J$
B	$\frac{R^2}{2(1+c^2)}J$	$\frac{R^2}{2}\left(1-\frac{c}{2}\right)J$	$\frac{R^2}{2}\left(1-\frac{c}{2}+0.998c\,\text{Kn}\right)J$
C	J	J	$(1+1.996\,\text{Kn})\,J$
D	$\frac{-c^2}{2R^2(1+c^2)}J$	$\frac{-c}{4R^2}J$	$\frac{-c}{4R^2}(1+1.996\,\text{Kn})\,J$
J	$U\left(-\frac{1}{2}\ln c - \frac{1}{2} + \frac{c^2}{1+c^2}\right)^{-1}$	$U\left(-\frac{1}{2}\ln c - \frac{3}{4} + c - \frac{c^2}{4}\right)^{-1}$	$U\left[\left(-\frac{1}{2}\ln c - \frac{3}{4} + c - \frac{c^2}{4}\right) + 1.996\,\text{Kn}\left(-\frac{1}{2}\ln c - \frac{1}{4} + \frac{c^2}{4}\right)\right]^{-1}$

The great strength of the single fibre theory is its simplicity, and for this reason it has been the most widely used theory in calculations of filter performance. Fitting the boundary conditions requires only the solution of four simultaneous equations, giving the constants in equation 3.31. The theory results in a simple analytical expression for the stream function, from which streamlines can be readily calculated, and using which calculations of particle capture efficiency can be easily carried out. A typical streamline pattern is shown in Fig. 3.7, for a fibre in a filter with a packing fraction of 0.05, calculated according to Kuwabara's model. The pattern according to Happel's theory is not dissimilar.

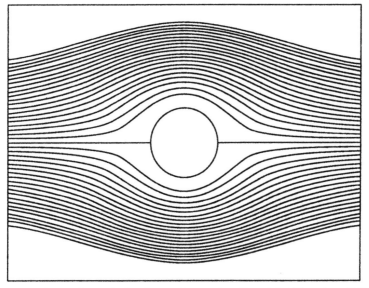

FIG. 3.7   Streamlines calculated according to the cell model of Kuwabara (19).

The flow pattern has upstream–downstream symmetry, and its shape does not vary with velocity. The streamlines for any velocity can be obtained simply by scaling the values of the stream function in Fig. 3.7. Experiments have been carried out (22) illustrating flow patterns by tracing, photographically, the trajectories of particles of neutral buoyancy suspended in a fluid. The flow pattern in model staggered arrays of fibres was very accurately described by both theories, but that of Kuwabara gave a better description of the flow velocity, and by implication the pressure drop.

Hydrodynamic factor

Much of filtration theory is concerned with the flow pattern close to the fibre surface. The form here can be obtained from equation 3.31 and Table 3.2 by making the substitution $r = R + \delta$ and considering only the lowest significant term in a Taylor expansion. The result, using the Kuwabara model, is similar in form to equation 3.9.

$$\psi \doteqdot \frac{U(1-c)\delta^2 \sin\theta}{RKu} \doteqdot \frac{U\,\delta^2 \sin\theta}{R\zeta} \qquad (3.34)$$

The constant Ku is identical to the bracketed expression in equation 3.33. The more general form in equation 3.34 holds for any model, with appropriate choice of the constant ζ. This means that in many instances a general theory may be developed, and the individual flow model identified by the value of the constant ζ, which is called the hydrodynamic factor (21).

The hydrodynamic factor calculated on the basis of this model, and those calculated from models to be studied below, are listed in Table 3.3. The hydrodynamic factor is related to the velocity gradient at the fibre surface and, therefore, to the viscous drag and to the pressure drop across the filter.

$$\Delta p = \frac{4\eta chU}{R^2\zeta} \qquad (3.35)$$

TABLE 3.3.
Hydrodynamic factors according to various theories

Lamb	$-\ln Re + 2.0002$
Kuwabara $(c\rightarrow 0)$	$-\frac{1}{2}\ln c - 0.75$
Happel $(c\rightarrow 0)$	$-\frac{1}{2}\ln c - 0.5$
Miyagi $(c\rightarrow 0)$	$-\frac{1}{2}\ln c - 0.76$
Hasimoto/Drummond–Tahir $(c\rightarrow 0)$	$-\frac{1}{2}\ln c - 0.74$
Spielman-Goren	$-\ln\left(\frac{\chi}{2}\right) - 0.5772$

N.B. There is basic consistency among the theories of Miyagi, Kuwabara, and Hasimoto (and Drummond and Tahir) in spite of the difference in the structures ascribed to the filters treated.

Flow parallel to the fibre axis, and at arbitrary inclination

The filter sections shown in Chapter 2 indicate that although fibres in a filter tend to be orientated roughly perpendicular to the airflow, this is not universally so, and any orientation may occur, even that in which the fibre axis is parallel to the flow. Fortunately the single fibre theories can easily be modified to describe this situation. Flow at any angle to the fibre can be described if the flow field is found in the two principal directions, parallel and perpendicular to the fibre. In flow parallel to the fibre the only non-zero component of velocity is U_z and this varies only with r. Symmetry requires that the pressure gradient is constant and that it acts in the z-direction only, and so equation 3.14 takes the form,

$$\frac{dp}{dz} = \eta \nabla^2 U_z.$$ (3.36)

One of the necessary boundary conditions for the solution of this equation is that U_z vanishes at the fibre surface. The remaining condition is applied at the cell surface, where the conditions of both zero tangential stress and zero vorticity become,

$$\frac{\partial U_z}{\partial r} = 0.$$ (3.37)

The result is that the air velocity is given by (18),

$$U_z = \frac{-Uc\left(r^2 - R^2 - \frac{2R^2}{c}\ln\left(\frac{r}{R}\right)\right)}{2R^2\left[-\frac{1}{2}\ln(c) - 0.75 + c - \frac{c^2}{4}\right]}$$ (3.38)

and the pressure drop by,

$$\Delta p = \frac{2\eta chU}{R^2\left[-\frac{1}{2}\ln(c) - 0.75 + c - \frac{c^2}{4}\right]}.$$ (3.39)

Equation 3.39 gives exactly one half of the value calculated by Kuwabara's model, and approximately one half of that by Happel's for flow perpendicular to the fibre axis. Flow at any angle relative to the fibre axis may be calculated as a sum of the components parallel and perpendicular to the fibres.

Refinement of single fibre theory to describe arrays of fibres

The single fibre theory described above takes account of only one aspect of filter structure, packing fraction. Moreover, though the flow pattern close to the fibre surface is likely to be well described by the theory, the description of flow in the interstitial regions will be poor; and since it is not possible completely to fill space with circular cylinders, cell theory cannot give a stream function that is continuous throughout a filter modelled as a regular array like Fig. 3.5.

A refinement to the theory is the inclusion of higher order solutions of the biharmonic equation (23, 24) which take the following form, for all values of n,

$$\psi_n = (A_n r^{2n-1} + B_n r^{1-2n} + C_n r^{3-2n}$$
$$+ D_n r^{2n+1}) \sin(2n-1)\theta. \tag{3.40}$$

The original solution, equation 3.31, plus a series of terms such as that in equation 3.40, each of which can satisfy the boundary conditions at the fibre surface, can be made to give a close fit to appropriate boundary conditions at a non-cylindrical outer surface, which could be chosen to have a shape that would be completely space-filling like prisms of rectangular or regular hexagonal cross-section. The resulting stream function would correspond to flow around fibres in arrays such as those shown in Fig. 3.5. Essentially similar calculations for flow parallel to the fibres have been carried out (25).

Continuum theory

A further theory of airflow around single fibres, but without the assumption of cells, is that of Spielman and Goren (26). Their solution is essentially a hybrid of macroscopic flow and viscosity-dominated flow close to the fibre, and they make the assumption that the two components can be added together to give the following modification to equation 3.14,

$$\nabla p = \eta \nabla^2 \underline{U} - \eta \underline{\underline{k}} \cdot \underline{U} \tag{3.41}$$

in which $\underline{k}$ is a tensor describing the macroscopic resistance of the filter material. The form of k depends on the arrangement of fibres within the filter, and the authors consider a number of different possibilities, including that in which the fibre arrangement is completely random and k, therefore, is a scalar.

The solution is more involved than that of the cell theories, and the pressure drop, for this model, is given by,

$$\Delta p = \frac{2\eta c h U}{R^2} \chi^{1/2} \frac{K_1(\chi)}{K_0(\chi)} \tag{3.42}$$

where K_0 and K_1 are modified Bessel functions, and χ is the solution of a transcendental equation which, for the case of a random arrangement of fibres, is,

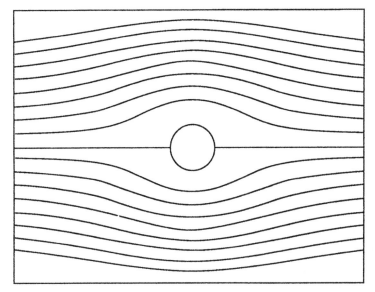

Fig. 3.8 Streamlines calculated according to the continuum theory of Spielman and Goren (26).

$$\chi^2 = 4c\left[\tfrac{1}{3}\chi^2 + \tfrac{5}{6}\chi\, \frac{K_1(\chi)}{K_0(\chi)} \right]. \qquad (3.43)$$

A typical flow pattern for this model is illustrated in Fig. 3.8.

Flow through two-dimensional arrays of fibres

Although the cell theory can be generalised to prismatic cells that can fill the whole of space, there also exist theories that take account of arrays of fibres at the outset. An analytical solution has been obtained, in the form of Fourier series (27) for the flow through a square array of cylinders. The author of the work does not give flow patterns, but calculates the drag force on a fibre of unit length, which can be related to the pressure drop using equation 3.23.

$$\Delta p = \frac{4\eta ch U}{R^2[-\tfrac{1}{2}\ln(c) - 0.738 + c + 0(c^2)]} \qquad (3.44)$$

To the order of precision quoted, this expression is identical with one obtained using matching methods including singularities, (28) by authors who solve the same problem for a range of symmetric arrays and for flow parallel and perpendicular to the fibre axes. The hydrodynamic factors, calculated in this way, along with values of the parameter obtained above, are summarised in Table 3.3.

These methods have the virtue of describing the pressure drop, in terms of the hydrodynamic factor, in an analytical form, but they are limited to fibres of circular cross section and to arrays of relatively high symmetry. Other theories have overcome these drawbacks.

Many-fibre theories

The flow pattern through and pressure drop across any array of fibres can be calculated by finite element or finite difference numerical methods (29, 30), but the calculations require a mesh to be set up throughout the array. The solution to the flow pattern is obtained in the form of information at the mesh points. Symmetry can be exploited so that this mesh is confined to the lowest repeated area of the array, i.e. that part which is large enough to be representative of the entire array; and these areas are indicated in Fig. 3.5. As the symmetry of the array is reduced, the irreducible area becomes larger.

Variational method

An alternative both to numerical methods and to the method of solving a flow equation is that of finding the stream function, expressed as an analytical function, that gives rise to the lowest rate of dissipation of energy by viscous drag (31), Θ, which depends explicitly on the stream function in the following way,

$$\Theta = \eta \int\int \left(\frac{\partial^2\psi}{\partial x^2} - \frac{\partial^2\psi}{\partial y^2}\right)^2 + 4\left(\frac{\partial^2\psi}{dx\,\partial y}\right)^2 \, dx\,dy. \qquad (3.45)$$

From the physical point of view, finding the minimum of equation 3.45 corresponds to Helmholtz's principle (32), that the flow pattern actually followed is that which satisfies the natural boundary conditions of the system, whilst giving rise to the lowest possible viscous dissipation; and from a mathematical point of view (33) equation 3.24, the biharmonic equation, is the Euler equation corresponding to equation 3.45.

A trial stream function containing adjustable parameters is required, and the optimal solution is given by those that give the lowest value of Θ. The trial function used is a Fourier series with the coefficients as adjustable parameters, and the orthogonality of the terms makes the integral in equation 3.45 particularly simple. The form of the series depends on the geometry of the array being modelled, and the simplest, corresponding to the channel array in Fig. 3.5, is,

$$\psi = Uy + Ub \sum_{n=1}^{\infty} \sum_{k=0}^{\infty} a_{nk} \sin \frac{n\pi y}{b} \cos \frac{k\pi x}{e}. \tag{3.46}$$

The boundary conditions to be satisfied are that the air in contact with the fibres should be stationary. The other conditions corresponding to the boundary conditions required for the single fibre theories are satisfied by the symmetry properties of the trial function. In practice the former conditions are included by requiring ψ to vanish at a finite number of points spaced around the fibre surface, in pairs, one member of each pair being just inside the surface and the other just outside it. If the coordinates of a general point where the condition is applied are (x_i, y_i), and if Θ is minimised subject to these constraints, introduced by means of Lagrangian multipliers (34) $\{\gamma_i\}$, the result, for a channel array of fibres is given by,

$$a_{nk} = \frac{b}{f_{nk}} \sum_{i=1}^{m} \gamma_i \sin \frac{n\pi y_i}{b} \cos \frac{k\pi x_i}{e} \tag{3.47}$$

where,

$$f_{nk} = 2\left(\frac{n^2}{b^2} + \frac{k^2}{e^2}\right)^2, \quad k \neq 0; \quad f_{n0} = \frac{4n^4}{b^4} \tag{3.48}$$

and the γ_i are the solutions of the simultaneous equations,

$$\sum_{i=1}^{m} \gamma_i \sum_{n=1}^{\infty} \sum_{k=0}^{\infty} \frac{b^2}{f_{nk}} \sin \frac{n\pi y_i}{b} \sin \frac{n\pi y_j}{b} \cos \frac{k\pi x_i}{e} \cos \frac{k\pi x_j}{e} = y_j.$$

$$\text{For all} \quad 1 \leq j \leq m. \tag{3.49}$$

The advantages of this model are that a continuous stream function throughout the filter is obtained in a simple form, that the flow pattern in the interstitial regions is realistic and that the model is not limited to fibres of circular cross-section (the last case is significant because filter materials exist with fibres of rectangular or irregular cross section). Furthermore, if a smallest "cell" is drawn for an array, there is no ambiguity about boundary conditions at the cell surface. The model can be applied to arrays with a lower degree of symmetry than the above, the simplest of these being the staggered array illustrated in Fig.

3.5. A drawback is that since the boundary conditions are applied at a finite number, M, of discrete points at the fibre surface, the model is at its poorest in this region.

The pressure drop through a channel model filter is given by equation 3.12 with,

$$f(c) = \frac{b^2 R^2 \pi^4}{4} \sum_{n=1}^{\infty} \sum_{k=0}^{\infty} a_{nk}^2 f_{nk} \qquad (3.50)$$

and account is taken of the structure by the dependence of f_{nk} on the lengths b and e. Calculations using this model, given in Table 3.1, predict that the pressure drop is not particularly sensitive to fibre arrangement, as commented elsewhere (28), provided that the nearest neighbour distance between fibres is kept more or less constant. The experimentally observed response of the pressure drop of a filter to compression (35), is accurately described by the model. The pattern of streamlines for airflow through a staggered array of fibres with circular cross-section (36) is shown in Fig. 3.9

FIG. 3.9   Streamlines calculated by the variational method, for an array of staggered geometry (31).

The method can be applied to flow at any orientation in three dimensions simply by adding velocity flow fields. It predicts that, unless the packing fraction is extremely high, the pressure drop when the flow is parallel to the fibre axes is approximately half of that when the flow is perpendicular, in basic agreement with other theories. In arrays that have square symmetry, the pressure drop does not vary with orientation perpendicular to the fibre axes, which is a two-dimensional example of a general physical principle: that properties depending on linear response are completely isotropic in systems of cubic symmetry.

Boundary element method

An alternative way of solving the problem of flow through arrays (37) employs the boundary element method (38) to obtain a solution of the biharmonic equation. The first step in this approach is to split the equation into two equations by way of the definition of vorticity.

$$\nabla^2 \psi = -\omega \tag{3.51}$$

$$\nabla^2 \omega = 0 \tag{3.52}$$

Each of these equations is then expressed in the form of an integral equation,

$$v(\underline{r})\psi(\underline{r}) = \int_c \psi(\underline{s})G_1'(\underline{r}, \underline{s}) - \psi'(\underline{s})G_1(\underline{r}, \underline{s})$$
$$+ \omega(\underline{s})G_2'(\underline{r}, \underline{s}) - \omega'(\underline{s})G_2(\underline{r}, \underline{s})ds \tag{3.53}$$

$$v(\underline{r})\omega(\underline{r}) = \int_c \omega(\underline{s})G_1'(\underline{r}, \underline{s}) - \omega'(\underline{s})G_1(\underline{r}, \underline{s})ds \tag{3.54}$$

where the contour, C, over which the integral is carried out, is the boundary of the region of interest, which is a repeating element of the fibre array. $\underline{r}$ and $\underline{s}$ are position vectors, and G_1 and G_2 are the fundamental free-space solutions of Laplace's equation and the biharmonic equation respectively,

$$G_1(\underline{r}, \underline{s}) = \ln|\underline{r} - \underline{s}| \tag{3.55}$$

$$G_2(\underline{r}, \underline{s}) = \tfrac{1}{4}|\underline{r} - \underline{s}|^2\{\ln|\underline{r} - \underline{s}| - 1\} \tag{3.56}$$

and v is a function that takes the value 2π within the region of interest, zero outside, and π on the border, except at sharp corners.

Four boundary conditions are required at each point on the boundary. Two of these follow from the physical properties of the system, and the other two are obtained by solving equations 3.53 and 3.54 at points on the boundary. Once the boundary conditions are known, equations 3.53 and 3.54 can be used to find the values of ψ at any point within the region of interest. The equations cannot be solved analytically for fibre arrays, and the procedure used is to divide the boundary of the region, both the fibre surface and the lines of symmetry, into discrete units, and to reduce the problem to a number of simultaneous equations. The results obtained using this method agree closely with those obtained by means of the variational method.

Filters of irregular and imperfect structure

The theory above has dealt with model structures. Understanding of the behaviour of real filters of imperfect structure can come about from empirical models, and from analytical approaches to well-defined imperfections.

Empirical models of pressure drop

Although the airflow pattern in a real filter defies simple measurement, the pressure drop can be measured easily. It has been shown that the structure of filters is complicated and variable; but it has also emerged from the study of the theory of airflow and pressure drop that the latter depends critically on the packing fraction, but more weakly on the fibre arrangement (a more detailed account of the effect of non-uniformity of structure will be given later in this chapter). In fact the pressure drop varies only by a factor of two between the instances where fibres are arranged parallel to and perpendicular to the airflow. It would be expected, then, that filters of common packing fraction would have reasonably similar pressure drops, even if no account is taken of structure; and this is observed. In practice correlations among a range of media (39) have been satisfactorily described by equation 3.12 along with the following empirical formula:

$$f(c) = 16c^{3/2}(1 + 56c^3). \tag{3.57}$$

A further, very extensive, study has been carried out (40) in which the measurements by a large number of authors have been plotted together.

The data, shown in Fig. 3.10, are all for low Reynolds number, and for low Knudsen number (*vid inf*) but are taken from a variety of physical systems, both liquids and gases. The values of packing fraction vary by four orders of magnitude and the values of $f(c)$ by almost six. It is clear from the figure that a strong pattern exists, though at the intermediate values, where most of the data points lie, there is a

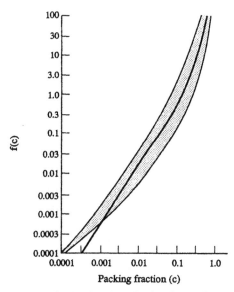

F$_{IG}$. 3.10   Envelope curves for $f(c)$ based on the data from reference 40. The bold line
is the empirical equation 3.57. (By permission of D. F. James.)

variation of an order of magnitude. The empirical equation 3.57 drawn
on the same graph gives a good description of typical results, except at
very low packing fraction.

As illustrated in Table 3.1, theoretical models predict a higher
pressure drop than experimental observations, two likely reasons being
that in real filters the fibres are not all perpendicular to the airflow, and
that in real filters the structure is not uniform.

The effect of fibres of non-circular cross-section

The many-fibre theory could be used to calculate the effect of any
chosen fibre cross-section on the performance of a filter, but two
examples of special interest will be considered here: fibres with small
surface asperities, and fibres with elliptical or rectangular cross-
sections.

Many thermo-plastic fibres have a relatively smooth surface, but
reconstituted cellulose fibres are furrowed, as are other fibres produced
by a wet spinning process; and wool fibres, which are widely used in air
filters, have a distinctly scaly surface. It is important to know whether
surface details such as these have a significant effect on airflow pattern.
There is a general principle (41) that the Stokes drag on a body is less
than or equal to that on a body that would completely enclose it and,
therefore, greater than or equal to that on one that it would enclose.
This means that the drag on a rough-surfaced fibre lies between that of

a smooth-surfaced fibre with a cross-section given by the inscribed circle, and that of a smooth-surfaced fibre with a cross-section given by the circumscribed circle. If surface asperities are small, these two circles will be almost identical and equal to the circle with the same area as the cross-section of the fibre, and so the effects of roughness on pressure drop would be small. Calculations using the many-fibre theory indicate that the effects on the flow pattern should also be small.

Fibres of elliptical cross section are important as approximations to the form of fibres that bear a deposit of captured material. Loaded filters will be treated in detail in a later chapter, but the theoretical models will be considered here. Because the cross-section still has a high degree of symmetry the problem can be solved by an extension of the Oseen approximation, the case chosen being a single layer of fibres at low but non-zero Reynolds number (42). Calculations of fibre drag and, therefore, filter pressure drop, were carried out for the case where the minor axis was one half of the major, for fibres that were five major axes apart, and the values compared with those for circular cylinder fibres with a diameter equal to that of the major axes. The fibres with major axis perpendicular to the flow had a drag of 84.5% of that of the cylindrical fibres, whereas those with axes parallel had a drag of only 48.9% of this value.

Numerical calculations of the drag on fibres of rectangular cross section were carried out (30) for the case where the packing fraction was kept constant and the aspect ratio varied. The drag was found to vary only little with aspect ratio up to 4, with the long axis of the rectangle aligned parallel to the flow; but when it was perpendicular, an aspect ratio of 6 caused the drag approximately to double at a packing fraction of 0.01, with a greater effect at higher packing fractions.

The form of an elliptical fibre that is the closest simple approximation to a fibre with a dust load is one with a constant minor axis, and with a major axis and, therefore, a cross-sectional area, that increases with the dust load. The flow through an array of such fibres can easily be solved by the variational method. Figure 3.11 shows the flow patterns and Table 4 shows the effect of the eccentricity of the ellipse on the calculated pressure drop of the filter. The effect is relatively small when the major axis of the ellipse is aligned along the flow direction, but larger when the two are perpendicular.

Pressure drop of filters made from polydisperse fibres

Filters that are made from carded textile fibres usually have fibres of approximately uniform size. However, paper filters frequently contain fibres with a wide range of sizes.

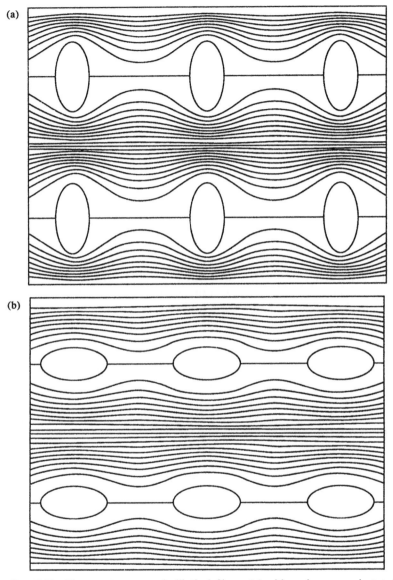

Fɪɢ. 3.11   Flow pattern around elliptical fibres: (a) with major aces orientated perpendicular to the flow; (b) with major axes orientated parallel to the flow (calculated according to the variational model).

An exact approach to the descriptions of these is not possible but a highly simplified model, consisting of a single layer of equally spaced fibres of alternating size, has been successfully studied both experimentally and theoretically (43). The flow field can be solved by means of the

TABLE 3.4
Effect of aspect ratio of fibre cross section on filter pressure drop, at constant minor axis

| | Scaled pressure drop | | | |
| | Minor axis $=1.0$ $b=e=3.0$ | | Minor axis $=1.0$ $b=e=5.0$ | |
Aspect ratio	Major axis parallel	Major axis perpendicular	Major axis parallel	Major axis perpendicular
1.0	1.00	1.00	1.00	1.00
1.1	1.02	1.13	1.02	1.07
1.2	1.04	1.27	1.04	1.15
1.5	1.10	1.92	1.09	1.42
2.0	1.17	4.63	1.16	2.06

complex variable (16) approach used previously and, as before, although the calculation is rather involved, the pressure drop emerges relatively easily. The value is close to that across a filter of monodisperse fibres with a diameter equal to the arithmetical mean of those of the fibres in question, provided that the ratio of sizes is less than 2–3. If the ratio is greater, the pressure drop is smaller than that calculated from the mean size. Agreement between theory and experiment was found to be reasonable for both the pressure drop and the flow pattern, illustrated by suspended tracer particles.

Experimental work on real glass fibre filters (44) made from mixtures of two different sizes of fibre has shown a linear dependence of pressure drop on the fractional mass of each component, which is a rather different relationship from that above. However, both point to the existence of relatively simple rules for the behaviour of filters made from fibre mixtures.

Flow in conditions other than simple Stokes flow

Although Stokes flow is the most successful simple approximation to airflow through filters, it frequently happens with filters of coarse fibres that the Reynolds number is too high for the condition to hold; and with filters of very fine fibres it may be necessary to take into account the molecular nature of the air.

Flow in fibre arrays at finite Reynolds number

The Reynolds number for coarse filters operated at relatively high velocity may be of the order of unity, making the solution of the flow problem for small but finite *Re* necessary. This can be effected using numerical methods such as the finite element method, but it can also be done for the case of layers of fibres by an extension of complex variable theory, and for that of a fibre array by an extension of the many fibre theory using the variational method.

The calculation of the pressure drop for a fibre layer (16) gives a first correction term of $O(Re^2)$. Theory and experiment are found to agree (45) as shown in Fig. 3.12. The pressure drop varies little with Reynolds number up to a value of 0.5, but at high values of Re the pressure drop increases considerably.

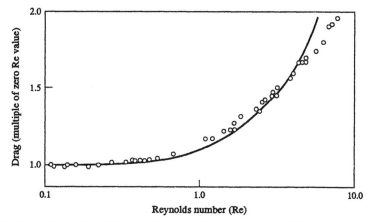

FIG. 3.12   Drag divided by zero Re value for a single layer of fibres.—Theory 0 Experiment (45). (With permission from Pergamon Press Ltd, Headington Hill Hall, Oxford OX3 0BW, U.K.).

The many-fibre theory can be generalised (46) by solving equation 3.16, which takes account of the inertia of the air in steady state conditions, rather than the simpler Stokes flow equation 3.24. The stream function for small but finite Re is then written as a series in an expansion parameter that is closely related to the Reynolds number.

$$\Psi = \sum_n \left(\frac{\rho_g U b}{\eta} \right)^n \psi_n \qquad (3.58)$$

The zero order term of this series is the Stokes flow term, which has already been calculated, and substitution of equation 3.58 into equation 3.16 and equating powers of Re enables an equation for ψ_1 to be obtained in terms of ψ_0. The use of Fourier series simplifies the matter, and by repeating the process, the higher order terms can be obtained. The flow pattern calculated by this means is shown in Fig. 3.13, which illustrates that the symmetry characteristic of Stokes flow

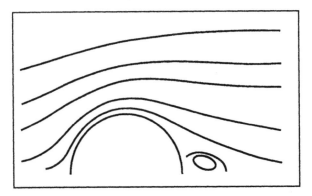

Fig. 3.13 Streamlines for flow at a Reynolds number of 3.76, through a channel array with a packing fraction of 0.031 (46). (© Crown Copyright; by permission of Her Majesty's Stationery Office.)

has been lost, and that the form of Ψ is velocity-dependent. The pressure drop, calculated in this way, is a series in odd powers of the flow velocity, the first term being the Stokes flow term and the next term a cubic. This functional form follows from general symmetry conditions applied to response theory, and a comparison between theory and experiment on model arrays is good, but not as good as that observed for a single layer of fibres.

A further feature that emerges is that though the first order term in pressure drop has only a weak structure dependence, the third order term is highly sensitive to structure. The results of theory (points) and experiment for two model filters of the same packing fraction but different structures (channel and staggered arrays), as shown in Fig. 3.14, illustrate this clearly.

Molecular nature of airflow through filters, and the effects of aerodynamic slip on flow pattern and pressure drop

In the theory developed so far it has been assumed that the air in contact with a fibre surface is stationary; and this gives rise to the boundary conditions applied at the fibre surface in airflow calculations. This assumption is, however, not strictly correct, for it does not take into account the molecular nature of the air. Gas molecules travel a finite distance before transfer, by collision, of momentum or any other property relevant to transport phenomena can come about. The mean distance, λ, is approximately 0.065 μm in air at NTP; and the effect of a finite mean free path in filters is that the velocity of air at a surface does not vanish in the tangential direction, but rather approaches a finite limit, the process being termed aerodynamic slip.

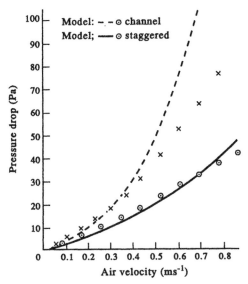

FIG. 3.14 Pressure drop as a function of velocity from the many fibre model: ⊙ ——, channel model; + − −, staggered model (46). (© Crown Copyright; by permission of Her Majesty's Stationery Office.)

General analysis

In the simplest analysis of the phenomenon of slip (47), attention is focused on the average tangential velocity of molecules one mean free path distant from the surface. The average is taken between those molecules that have been reflected from the surface and those that are approaching it, assumed to be equal in number. The former velocity depends on the nature of reflection at the surface; if reflection is specular the tangential velocity will equal that of the molecules at their approach; if diffuse it will equal that of the surface itself. The velocity of particles approaching will be related to the velocity gradient within the volume of the air, which is governed by the equations of motion of the air, assumed to be unaltered in this region. The result of the analysis is an expression including the velocity gradient and the mean free path. In the simplest system, that of Couette flow, with two planar parallel surfaces, separated by a distance d, moving at constant relative velocity, and with completely diffuse reflection of molecules on impact, the velocity gradient becomes,

$$\frac{dU}{dz} = \frac{U}{d}\left(1 + \frac{2\lambda}{d}\right)^{-1} \tag{3.59}$$

the second term in brackets being a dimensionless parameter

quantifying the effect of slip. The expression reverts to its simple classical form as $\lambda \to 0$.

Aerodynamic slip applied to fibres

For flow around a filter fibre (48), the tangential velocity at the fibre surface when aerodynamic slip occurs, is not zero, but is related to the tangential stress, as defined in equation 3.19,

$$U_\theta = \xi \left(\frac{\partial U_\theta}{\partial r} + \frac{1}{r} \frac{\partial U_r}{\partial \theta} - \frac{U_\theta}{r} \right)_{r=R} \tag{3.60}$$

ξ is a constant, the form of which depends on the nature of the reflection of molecules at the fibre surface. If the reflection is entirely diffuse,

$$\xi = 0.998 \; Kn \tag{3.61}$$

where Kn, the Knudsen number, is given by,

$$Kn = \frac{\lambda}{R}. \tag{3.62}$$

Three regimes (49) where molecular effects are important can be defined: the slip regime ($Kn \leq 0.25$), in which case the preceding theory applies; the free molecular regime ($Kn \geq 10$), in which valid approximations based on free molecular theory can be used; and a transitional regime between these two, in which no analytical approximations apply. At normal pressures the free molecular regime would apply only in situations where the fibre diameters are less than 0.01 μm, and in later chapters only the slip regime will be studied in detail.

The complex variable theory (16) can be modified (50) so that the tangential velocity condition applied at the fibre surface takes slip into account. Since the drag acting on a fibre can be expressed as a function of only the leading term of the velocity series expansion, the calculation is relatively straight-forward. The resulting expression for the dimensionless drag is,

$$\frac{1}{F_d(Kn)} = \frac{1}{F_d(o)} + \frac{\tau}{4\pi} \left[1 - \frac{1}{3} \left(\frac{\pi R}{2b} \right)^2 \right] Kn \tag{3.63}$$

where τ is a parameter of the order of unity, the exact value of which depends on the nature of the reflection of molecules at the fibre surface. The second term in brackets is equal to the packing fraction of a filter, if that filter were made from stacks of fibre layers separated by the inter-fibre distance.

The authors (51) of the work above carried out experiments, at reduced pressures and therefore increased Kn, on filters of various structures, which validated the form of equation 3.63. They found that filters with the same fibre diameter and packing fraction but different structures performed in different ways; and they were able to include the structures by way of an index, ε, which is the quotient of the resistance of the filter under consideration and that of a fan model filter with the same fibre diameter and packing fraction. This index could be included empirically into equation 3.63 to give the result, consistent with earlier measurements (52), carried out at reduced pressures.

$$\frac{1}{F_d(Kn)} = \frac{1}{F_d(o)} + \frac{\tau\varepsilon^{1/2}}{4\pi}(1-c)Kn \tag{3.64}$$

Cell models can also be generalised to account for slip, since equation 3.24 is still easy to solve, even with this slightly more complicated boundary condition; and its solution, for the Kuwabara flow field, gives an expression with the form of equation 3.31, but with modified constants, as shown in Table 3.2. The calculated pressure drop, on the basis of the Kuwabara field with aerodynamic slip, given by the treatment outlined above, is,

$$\Delta p = \frac{4\eta chU(1+1.996Kn)}{R^2\left[-\frac{1}{2}\ln(c)-0.75+c-\frac{c^2}{4}+1.996Kn\left(-\frac{1}{2}\ln(c)-0.25+\frac{c^2}{4}\right)\right]}. \tag{3.65}$$

The effect of aerodynamic slip on flow pattern is shown in Fig. 3.15, where the streamlines for flow around a fibre are calculated for $Kn=0.25$, which is considered to be the upper limit of Knudsen number for which the slip approximation is valid (49). It can be seen that the streamlines pass closer to the fibre when aerodynamic slip takes place; and the pressure drop when this occurs is invariably less than it would be otherwise.

Pich's expression (53) for pressure drop in the regime of free molecular flow is included for the sake of completeness;

$$\Delta p = \frac{2.29\eta chU}{R\lambda}. \tag{3.66}$$

The variation of pressure drop with Knudsen number for several

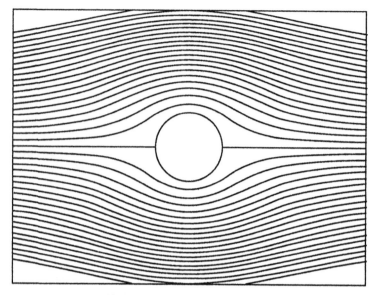

FIG. 3.15 Streamlines for Kuwabara flow with a packing fraction of 0.05; with $Kn = 0.25$, illustrating streamlines passing close to the fibre surface.

filters made from ultra-fine fibres and operated at a variety of absolute pressures has been clearly demonstrated (54). Though the experimental results were not compared directly with theoretical predictions, the general shape of the data was consistent with equation 3.64. Moreover the authors showed that at small Reynolds number the relative effect of increasing Kn was independent of the face velocity, as predicted.

The effect of aerodynamic slip on a filter with polydisperse fibres has also been calculated (51), and found to be described by an approximation of the same sort as that used above (43). The pressure drop is given by an expression identical to equation 3.64 except that the Knudsen number, Kn, is replaced by a similar dimensionless parameter calculated on the basis of the arithmetical mean of the fibre diameters.

Macroscopic flow patterns

Flow on a scale small compared with the dimensions of the filter, but large compared with the fibres is important in its own right, and is simple in concept. Darcy (4) observed that the pressure drop across filters is directly proportional to the rate of fluid flow through them, which has been shown in equation 3.12 to be a result of Stokes flow on a microscopic level. The fluid velocity for flow obeying Darcy's Law satisfies the equation,

$$U = -\frac{1}{\eta k}\nabla p \tag{3.67}$$

where k, the constant quantifying the resistance of the filter, is identical with the constant in equation 3.41.

Equation 3.67 shows that the airflow velocity is the gradient of a scalar potential (pressure), and is, therefore, potential flow (55). The curl of a gradient is zero, and this means that the vorticity, which is the curl of the velocity, is zero in this type of flow; it is irrotational. This can be readily confirmed by combining equations 3.2, 3.18 and 3.67. That the flow should be irrotational on this scale is reasonable, since rotational motion on a scale larger than that of the interstitial spaces would be destroyed as the air passes through the filter. On a smaller scale this does not apply. Equation 3.67 applies strictly only if the filter material is homogeneous and isotropic; and when the equation holds it means that macroscopic flow patterns through filters can be obtained from the potential flow equations, or by the use of analogues such as electrolyte baths or conducting paper. Macroscopic flow of this sort has, like Stokes flow, the property that the flow pattern does not alter with velocity. In addition, although the pressure drop will vary with the resistance of the medium, the flow pattern will not.

Flow through pleats

In some situations macroscopic flow is effectively two-dimensional, an example being airflow through pleated material, where the flow does not vary along the direction of the creases of the pleats. In such two-dimensional potential flow, the stream function obeys Laplace's equation,

$$\nabla^2 \psi = 0. \tag{3.68}$$

It is easier to tackle this particular problem by minimising energy dissipation than by solving the equation directly. Since the rate of dissipation of energy is equal to the product of the air velocity and the pressure drop, equation 3.67 can be used to show that this rate, Θ, is given in two-dimensional flow, by,

$$\Theta = \eta k \int \int \left(\frac{\partial \psi}{\partial x}\right)^2 + \left(\frac{\partial \psi}{\partial y}\right)^2 dx\, dy \tag{3.69}$$

and just as the biharmonic equation 3.24 is the Euler equation of the functional in equation 3.45 for microscopic flow, so is Laplace's equation the Euler equation of the functional in equation 3.69. Figure

3.16 shows the flow pattern through shallow pleats calculated by minimising this functional (56).

Pleating increases the area of the material that can be accommodated in a fixed volume and so reduces the filtration velocity and, therefore, the pressure drop at fixed volume flow rate. The pressure drop should, therefore, decrease as the number of pleats per unit length is increased; but eventually the restricted space between pleats will cause the pressure drop to rise again. Experimental results for this effect are rare, but one set is shown in Fig. 3.17. An optimal degree of pleating is clear, but the results do not follow any simple theoretical pattern (57). The ideal number of pleats may vary with the material used and the conditions of filtration.

Flow through filters of non-uniform structure

The sections of filters in Chapter 2 show that neither a regular array of fibres on a microscopic level, nor a homogeneous continuum at a macroscopic level gives a true picture of filter structure. In a study of the properties of a filter of non-ideal structure, deviations at two levels will be considered: those in which the relationship between fibres and their near neighbours is altered, but in which all parts of the filter have the same degree of alteration; and those in which different regions, each containing many fibres, have different densities and resistances.

Non-ideal structure on a microscopic level

In the many-fibre theory studied so far, although a range of structural variations were considered, the distance between each fibre and its nearest neighbours does not vary greatly from one structure to another. An experimental study of model filters with parallel fibres (35) in which the fibres are grouped together in pairs, as illustrated in Fig. 3.18, showed that this structural feature could affect pressure drop. If the alteration is such that the ratio of successive distances between fibres in the same layer is 3:1, the pressure drop is reduced to 72% of its previous value for filters with a packing fraction of 0.135 and to 90% for filters with a packing fraction of 0.060. The difference would be less for filters with a lower packing fraction still, but in every instance the effect on pressure drop should be a reduction.

The effect of rotating successive layers relative to one another, to produce a "Fan Model" filter, has also been measured. At a packing fraction of 0.05 the resistance of the Fan Model filters is reduced to 82% of that of the staggered geometry model.

Fibre–fibre contact is important (58) because, as stated in Chapter 2, although such contacts are relatively rare, without them a physically

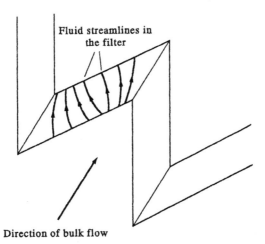

Fluid streamlines in
the filter

Direction of bulk flow

FIG. 3.16 Calculated flow pattern through a shallow pleat (56). (© Crown
Copyright; by permission of Her Majesty's Stationery Office.)

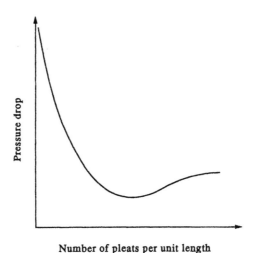

Number of pleats per unit length

FIG. 3.17 Scaled pressure drop through a pleated filter as a function of the number
of pleats (57). (With permission from Elsevier Science Publishers.)

stable filter could not exist. By analogy with the work above, contact
between roughly parallel fibres would be expected to reduce the
pressure drop. If the contacting fibres are perpendicular, the leading
fibre will create a region of relative stagnation and, therefore, reduce
the average air velocity past the following fibre, again reducing the
pressure drop across the filter.

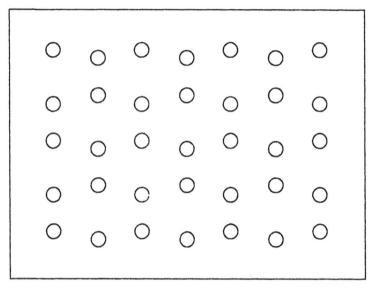

FIG. 3.18   Structure of filter with microscopic imperfection taking the form of paired fibres.

Non-ideal structure on a macroscopic level

In principle the flow pattern through a real filter of non-uniform structure could be calculated by finite element or finite difference methods, but the grid required would have to extend throughout the entire filter since no symmetry simplifications could be used. To carry out such a calculation would require a great deal of computer time, and it would be difficult to know what could be done with the solution once it had been obtained.

For a feasible solution, study must be made at an intermediate scale, taking account of the variation of the resistance of the material with position. The calculation is not possible unless a quantitative knowledge of this permeability is obtained from experiment, or from calculations on the basis of microscopic theory.

Filters heterogeneous in directions perpendicular to the flow

If a filter is non-uniform in structure in the direction of flow only, it can be considered to be in every respect equivalent to a number of homogeneous filters in series. If the filter is non-uniform only in the direction perpendicular to the flow the problem is only slightly more complicated. Such a filter can be considered to be a number of homogeneous filters of different resistances in parallel; and the nature

of the flow is such that the pressure drops across all filter elements are the same. The result is that to the lowest order,

$$\Delta p = \Delta p_0 \left[1 - \frac{\sigma_c^2}{2} f(c) \frac{d^2}{dc^2} \left(\frac{1}{f(c)} \right) \right] \qquad (3.70)$$

where Δp_0 is the pressure drop that would be experienced, at the same flow rate, by a uniform filter with a resistance equal to the mean resistances of the filter elements, and σ_c is the coefficient of variation of the packing fraction. The resistance of the heterogeneous filter is invariably less than that of the homogeneous.

The variation has been quantified (59) in a theoretical model in which the filter is divided into sections parallel to the flow direction, each section having a random number of fibres, except that zero is not allowed. The resistance of each section is calculated on the basis of the fibre density using an accepted uniform filter model. The total resistance obtained depends on the scale chosen, and if a filter is divided into 25 such segments the resistance is about one half to two thirds that of a uniform filter of the same density.

Filters heterogeneous in directions parallel and perpendicular to the flow

If a filter is non-uniform in two directions the problem of calculating the airflow pattern is considerably more difficult, but there are, at present, two possible approaches, both of which require a local resistance to be definable.

The filter can be split up into discrete elements: squares in a two-dimensional analysis and cubes in a three-dimensional (60). Each element is assigned the resistance of a filter with the same packing fraction, calculated on the basis of a simple model, the Kuwabara model in the example chosen. If the resistance is written as in equation 3.67, but as a function of position, Darcy's law and the equation of continuity can be combined to give,

$$\nabla \cdot \left(\frac{1}{\eta k} \nabla p \right) = 0. \qquad (3.71)$$

This equation is then solved on the mesh of elements, the pressure drop on each being related to that at its four nearest neighbours in the two-dimensional case and six in the three-dimensional. The problem is reduced to the solution of linear simultaneous equations; and streamlines follow from interpolation of the results. A typical result is shown in Fig. 3.19.

The resistance of a heterogeneous filter, calculated in this way, is invariably lower than that of a homogeneous filter of the same average density. If the elements are assumed to have permeabilities of a random Gaussian distribution, the quotient of permeabilities resembles a quadratic function of the coefficient of variation of the density.

An alternative method (61), which does not require discrete elements, but which is limited to two-dimensional flow, uses a local density modification to the functional in equation 3.69. With essentially the same local density approximation, the rate of dissipation of energy may be written,

$$\Theta = \eta \int \int k(x, y) \left\{ \left(\frac{\partial \psi}{\partial x} \right)^2 + \left(\frac{\partial \psi}{\partial y} \right)^2 \right\} dx \, dy. \qquad (3.72)$$

This functional can be minimised, but there is no simple corresponding Euler equation.

Flow through pin holes and leaks

Pinholes and leaks can be considered to be extreme cases of heterogeneity of filters. In Stokes flow conditions, the rate of airflow through a pinhole, approximately by a capillary of length equal to the filter thickness, is given by the well-known Poiseuille formula (1).

$$Q_V \sim \frac{\Delta p \, d^4}{\eta l} \qquad (3.73)$$

Since the pressure drop across the leak will be the same as that across the filter, the extent of leakage can be simply calculated if it is assumed that Stokes flow takes place through the leaks, though a correction may be necessary for entry effects.

At high flow rates the situation may be more complicated. A pinhole or leak must have dimensions significantly larger than the interstitial spaces in the filter for it to be recognised as a leak at all, and this means that the Reynolds number (equation 3.1) relative to the leak is larger than that relative to the filter as a whole. As a result Stokes flow conditions break down at a lower air velocity. Since leaks invariably get more than their share of the airflow, Stokes flow may cease at a relatively low volume flow rate; and when this occurs the linear relationship between Q_V and Δp for the leak is replaced by,

$$Q_V \sim \Delta p^\alpha \qquad (3.74)$$

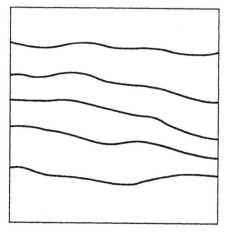

FIG. 3.19 Typical streamlines calculated for a heterogeneous filter from the incremental model (60). (With permission from Springer-Verlag, Heidelberg.)

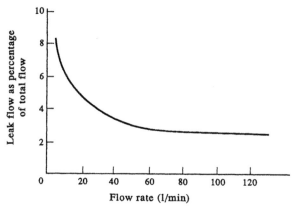

FIG. 3.20 Fraction of air passing through a leak, as a function of total flow through the filter (60). (By permission of the American Industrial Hygiene Association.)

where the index $\alpha \to 0.5$ at high airflow. The effect of this (62, 63) is that the fraction of air passing through the leak or pinhole decreases with increasing flow rate, as illustrated in Fig. 3.20.

Flow through a pinhole in a filter is different in a further respect from flow through an isolated pinhole, because in the latter case the velocity drops to zero at the solid pinhole boundaries whereas in a filter the velocity at the pinhole edge will match the finite velocity of air through the filter material. In addition to this there will be a boundary layer within the filter (64), as shown in Fig. 3.21, where the velocity exceeds the mean macroscopic velocity through the filter.

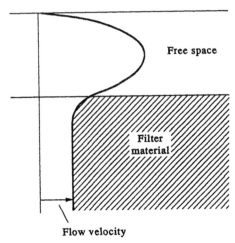

Free space

Filter material

Flow velocity

FIG. 3.21   Velocity profile through a channel bordering porous media, illustrating relaxation of the boundary condition of zero flow at the channel edge, and increased flow velocity at the filter edge (64). (By permission of W. K. Nader.)

References

1. NEWMAN, F. H. and SEARLE, V. H. L. *The General Properties of Matter*, p. 405, Edward Arnold, London, 1963.
2. LANGLOIS, W. E. *Slow Viscous Flow*, Macmillan, New York, 1964.
3. HUNTLEY, H. E. *Dimensional Analysis*, Macmillan, London, 1952.
4. DARCY, H. P. *Les Fontaines Publiques de la Ville de Dijon*, Dalaman, 1856.
5. LANDAU, L. D. and LIFSHITZ, E. M. *Fluid mechanics*, Pergamon, London.
6. HAPPEL, J. and BRENNER, H. *Low Reynolds Number Hydrodynamics with Special Reference to Particulate Media*, Prentice Hall, 1965.
7. OSEEN, C. W. Ark. Mat. Astron. Fiz., 1910, **6**(29) (see also H5 ch4, Ref. 10 Chap. 4).
8. LAMB, H. Motion of a sphere through a viscous fluid. *Philos. Mag.*, 1911, **21**, 112–121.
9. DAVIES, C. N. Viscous flow transverse to a circular cylinder. *Proc. Phys. Soc.*, 1950, **B63**, 288–296.
10. TOMOTIKA, S. and AOI, T. The steady flow of a viscous fluid past a sphere and circular cylinder at small Reynolds number. *Q. J. Mech. Appl. Math.*, 1950, **3**, 140–161.
11. BAUMGARTNER, H., LOEFFLER, F. and UMHAUER, H. Deep-bed electret filters—the determination of single fibre charge state and collection efficiency. *IEEE Trans. Electr. Insul.*, 1985, 477–486.
12. KOZENY, Ber. Wien. Akad., 1927, **136a**, 271.
13. CARMAN, P. C. Fluid flow through granular beds. *Trans. Inst. Chem. Eng.*, 1937, **15**, 150–166.
14. TIEN, C. *Granular Filtration of Aerosols and Hydrosols*, Butterworths, Boston, 1989.
15. TAMADA, K. and FUJIKAWA, H. The steady two-dimensional flow of viscous fluid at low Reynolds numbers passing through an infinite row of equal parallel circular cylinders. *Q. J. Mech. Appl. Math.*, 1957, **X**(4), 425–432.
16. MIYAGI, T. Viscous flow at low Reynolds numbers past an infinite row of equal circular cylinders. *J. Phys. Soc.*, Japan, 1958, **13**(5), 493–496.
17. KIRSCH, A. A., STECHKINA, I. B. and FUCHS, N. A. Effect of gas slip on the pressure drop in a system of parallel cylinders at small Reynolds numbers. *J. Coll. Interface Sci.*, 1971, **37**(2), 458–461.
18. HAPPEL, J. Viscous flow relative to arrays of cylinders. *AIChE J*, 1959, **5**(2), 174–177.

19. KUWABARA, S. The forces experienced by randomly distributed parallel circular cylinders or spheres in a viscous flow at small Reynolds numbers. *J. Phys. Soc., Japan*, 1959, **14**(4), 527–532.

20. IMAI, I. On the asymptotic behaviour of viscous fluid flow at a great distance from a cylindrical body, with special reference to Filon's paradox. *Proc. Roy. Soc.*, 1951, **A 208**, 487–516.

21. PICH, J. In *Aerosol Science* (ed. C. N. DAVIES), Academic Press, London, 1964.

22. KIRSCH, A. A. and FUSCH, N. A. The fluid flow in a system of parallel cylinders perpendicular to the flow direction at small Reynolds number. *J. Phys. Soc., Japan*, 1967, **22**(5), 1251–1255.

23. GOLOVIN, A. M. and LOPATIN, V. A. Flow of a viscous fluid through a doubly periodic series of cylinders. *Prikl. Mech. Teknich. Fiz.*, 1969, **2**, 99–105 (in Russian).

24. SANGANI, A. S. and ACRIVOS, A. Slow flow past periodic arrays of cylinders with application to heat transfer. *Int. J. Multiphase Flow*, 1982, **8**(3), 193–206.

25. SPARROW, E. M. and LOEFFLER, A. L. Longitudinal laminar flow between circular cylinders arranged in regular array. *AIChE J*, 1959, **5**(3), 325–330.

26. SPIELMAN, L. and GOREN, S. L. Model for predicting pressure drop and filtration efficiency in fibrous media. *Environ. Sci. Technol.*, 1969, **2**(4), 279–287.

27. HASIMOTO, H. On the periodic fundamental solutions of the Stokes flow equations and their application to viscous flow past a cubic array of spheres. *J. Fluid Mech.*, 1959, **5**, 317–328.

28. DRUMMOND, J. E. and TAHIR, M. I. Laminar viscous flow through regular arrays of parallel circular cylinders. *Int. J. Multiphase Flow*, 1984, **10**, 515–540.

29. PANTANKAR, S. V. *Numerical Heat Transfer and Fluid Flow*, McGraw Hill, New York, 1980.

30. FARDI, B. and LIU, B. Y. H. Flow field and pressure drop of filters with rectangular fibres. *Aerosol Sci. Technol.*, 1992, **17**, 36–44.

31. BROWN, R. C. A many fibre model of airflow through a fibrous filter. *J. Aerosol Sci.*, 1984, **15**(5), 583–593.

32. HELMHOLTZ, H. L. F. *Verh. Naturforsch. Med. Ser.*, 1886, 5.

33. BURLEY, D. M. *Studies in Optimization*, Galliard, Great Yarmouth.

34. RUSHBROOKE, G. S. *Introduction to Statistical Mechanics*, Clarendon, Oxford.

35. KIRSCH, A. A. and FUCHS, N. A. Studies on fibrous aerosol filters—11. Pressure drops in systems of parallel cylinders. *Ann. Occ. Hyg.*, 1967, **10**, 23–30.

36. BROWN, R. C. A model of filter behaviour including the effects of fibre shape and filter structure. *International Symposium on Air Pollution Abatement by Air Filtration and Related Methods* (*Respiratory Protection*), Dantest Copenhagen, 1985.

37. HILDYARD, M. L., INGHAM, D. B., HEGGS, P. J. and KELMANSON, M. A. Integral equation solution of viscous flow through a fibrous filter. In *Boundary Elements VII* (ed. C. A. BREBBIA and G. MAIER), Springer-Verlag, Berlin, 1985.

38. BREBBIA, C. A., TELLES, J. C. F. and WROBEL, L. C. *Boundary Element Techniques: Theory and Applications in Engineering*, Springer-Verlag, Berlin and New York, 1984.

39. DAVIES, C. N. The separation of airborne dust and particles. *Proc. Inst. Mech. Eng. (London)*, 1952, **B1**, 185–213.

40. JACKSON, G. W. and JAMES, D. F. The permeability of fibrous porous media. *Can. J. Chem. Eng.*, 1986, **64**, 364–374.

41. HILL, R. and POWER, G. Extremum principles for slow viscous flow and the approximate calculation of drag. *Q. J. Mech. Appl. Math.*, 1956, **9**, 313–319.

42. KUWABARA, S. The forces experienced by a lattice of elliptic cylinders in a uniform flow at small Reynolds number. *J. Phys. Soc., Japan*, 1959, **14**(4), 522–527.

43. KIRSCH, A. A. and STECHKINA, I. B. Pressure drop and diffusional deposition of aerosol in polydisperse model filters. *J. Coll. Int. Sci.*, 1973, **43**(1), 10–16.

44. STENHOUSE, J. I. T. Filters of polydisperse fibres. *Aerosol Society Conference*, 1990.

45. KIRSCH, A. A. and STECHKINA, I. B. Inertial deposition of aerosol particles in model filters at low Reynolds numbers. *J. Aerosol Sci.*, 1977, **8**, 301–307.

46. BROWN, R. C. Many fibre theory of airflow through a fibrous filter—II: Fluid inertia and fibre proximity. *J. Aerosol Sci.*, 1986, **17**(4), 685–697.

47. CHAPMAN, S. and COWLING, T. G. *The Mathematical Theory of Non-uniform Gases*, C. U. P., Cambridge, 1953.

48. PICH, J. The pressure drop of fibrous filters at small Knudsen numbers. *Ann. Occ. Hyg.*, 1966, **9**, 23–27.

49. PICH, J. Pressure characteristics of fibrous aerosol filters. *J. Coll. Int. Sci.*, 1971, **37**(4), 912–917.

50. KIRSCH, A. A., STECHKINA, I. B. and FUCHS, N. A. Effect of gas slip on the pressure drop in fibrous filters. *Aerosol Science*, 1973, **4**, 287–293.

51. KIRSCH, A. A., STECHKINA, I. B. and FUCHS, N. A. Gas flow in aerosol filters made of polydisperse fibres. *Aerosol Science*, 1974, **5**, 39–45.

52. STERN, S. C., ZELLER, H. L. and SHECKMAN, A. I. The aerosol efficiency and pressure drop of a fibrous filter at reduced pressures. *J. Coll. Sci.*, 1960, **15**, 546–562.

53. PICH, J. Pressure drop in fabric filters in molecular flow. *Staub Reinhalt. Luft.*, 1969, **29**(10), 10–11 (in English).

54. ZHANG, Z. and LIU, B. Y. H. Aerosol filtration in the transition flow regime. *Aerosol Sci. Technol.*, 1992, **16**, 227–235.

55. BATCHELOR, G. K. *Introduction to Fluid Dynamics*, C. U. P., Cambridge, 1967.

56. BROWN, R. C. The use of the variational principle in the solution of Stokes flow problems in fibrous filters. *J. Phys. D.*, 1983, **16**, 743–754.

57. DEAN, J. H. Non-woven wet-laid filter media. *Filtr. Sep.*, Nov/Dec **1972**, 669–674.

58. MONSON, D. R. Key parameters used in modelling pressure loss of fibrous filter media. In *Fluid Filtration. Gas*, Volume 1 (ed. R. R. RABER), American Society for Testing and Materials, Philadelphia, 1986.

59. YU, C. P. and SOONG, T. T. A random cell model for pressure drop prediction in fibrous filters. *J. Appl. Mech., Transactions of the ASME*, June 1975, 301–304.

60. LAJOS, T. The effect of inhomogeneity on flow in fibrous filters. *Staub Reinhalt. Luft.*, 1985, **45**(1), 19–22 (in English).

61. BROWN, R. C. Air flow through filters at a semimicroscopic scale (to be published).

62. FAHRBACH, J. The effect of leaks on the total penetration velocity and concentration measurements at perforated filters. *Staub Reinhalt. Luft.*, 1970, **30**(12), 45–52 (in English).

63. HINDS, W. C. and KRASKE, G. Performance of dust respirators with facial seal leaks 1. Experimental. *Am. Ind. Hyg. Assoc. J.*, 1987, **48**(10), 836–841.

64. NEALE, G. and NADER, W. Practical significance of Brinkman's extension of Darcy's law. Coupled parallel flows within a channel and a bounding medium. *Can. J. Chem. Eng.*, 1974, **52**, 475–478.

CHAPTER 4

Particle Capture by Mechanical Means

Introduction

Mechanical capture mechanisms are mechanisms that come into effect without the influence of attractive forces between the airborne particles and the filter fibre. They comprise: direct interception, which involves a particle following a streamline and being captured if this results in it coming into contact with the fibre; inertial impaction, in which capture is effected by the deviation of a particle from a streamline because of its own inertia; diffusional deposition, in which the combined action of airflow and Brownian motion brings a particle into contact with a fibre; and gravitational settling, the mechanism of which requires no description.

   This separation of effects is, of course, artificial because all will act simultaneously even if one in particular is dominant. Situations in which only one effect needs to be considered are easiest to deal with, and much of the chapter will be devoted to these. In the working that follows, single fibre efficiency, E, and dimensionless parameters, N, will feature significantly, and each will be subscripted according to the process that it relates to. A simple approach to the topic will be given at the beginning of each section, but first a generalisation of the definition of single fibre efficiency, given in Chapter 1, is necessary.

Generalised theory of single fibre efficiency

Single fibre efficiency, defined in Chapter 1, has the value,

$$E_s = \frac{2y}{d_f} \tag{4.1}$$

where the parameters y and d_f are as illustrated in Fig. 4.1. This definition is satisfactory provided that the fibre under consideration is isolated or almost isolated, and that a limiting trajectory can be defined in such a way that particles originating nearer to the axis than this trajectory will be captured and those originating further away will not. The first condition is violated if the fibre under question is preceded by other fibres and the second if the particles undergo random motion. In these situations the single fibre efficiency must be defined in one of two alternative ways.

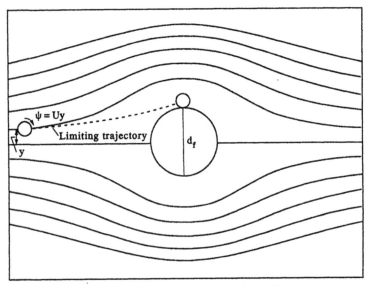

FIG. 4.1   Alternative definition of single fibre efficiency.

If the limiting particle can be considered to originate from a particular streamline with stream function ψ, and if the stream function at the fibre surface has a value of zero, the single fibre efficiency can be defined as,

$$E_s = \frac{2\psi}{U\,d_f}.$$

(4.2)

The value of ψ would, in Fig. 4.1, be Uy.

Alternatively, if a limiting trajectory cannot be identified, as in the case of capture by diffusional deposition, the single fibre efficiency can be related to the concentration of particles in the air, N, the velocity of approach, U, and the rate of capture of particles per unit time per unit length of fibre, n,

$$E_s = \frac{n}{NU\,d_f}.$$ (4.3)

Single fibre efficiency, however it is defined, is dimensionless. Relationships between it and other quantities must contain the latter in the form of dimensionless parameters as well; and specific dimensionless parameters exist for each capture process.

Limitations to the single fibre approach

The idea of single fibre efficiency is based on the assumption that all fibres in the filter can be treated as identical. However, even if the fibres have identical diameters, their environments will vary considerably in a filter with the irregular structure shown in the filter sections of Chapter 2. The correct approach to such a structure is to carry out a number of calculations of behaviour, corresponding to all environments, and then to average the results. This, however, is not usually feasible, and the approximation used in practice is to average the environment of a fibre first and then to carry out a single calculation of its behaviour. The approach is inexact, but it does lead to a usable average and it is the approach used throughout this book.

Particle capture by direct interception

Direct interception occurs when airborne particles behave in an entirely passive way with respect to the airflow; i.e. when they are subject neither to inertial effects nor to diffusive motion, and they are not acted upon by any external forces, including gravity. Any particles carried by the air in this situation would follow a streamline, as illustrated in Fig. 4.2; the value of the stream function at the position of the particle would not alter as the particle moved. In reality particles of finite size would be influenced by shear forces caused by the velocity gradient of the air across the particle dimensions, and very large particles would themselves influence the airflow. Complications like this would require the solution of a time-dependent flow equation, but provided that the particles are small compared with the dimensions of the fibres and the inter-fibre spacings, the complications can be neglected.

Simplified approach to direct interception

The capture of particles by interception can be described entirely in terms of the airflow itself, and this enables us to make some generalisations about the nature of this type of capture. The airflow

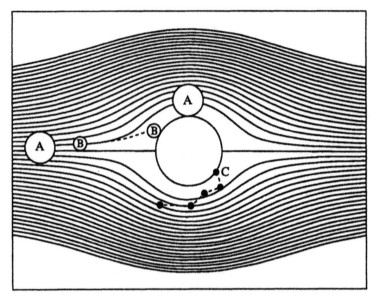

FIG. 4.2 Particle capture mechanisms: A, particle capture by interception; B, particle captured by inertial impaction; C, particle captured by diffusional deposition.

pattern in Stokes flow is independent of velocity and so it follows that capture by interception must also be velocity-independent. The particles do not move relative to the air and so air viscosity can have no effect. Air density likewise has no influence, except that at low pressures and therefore low densities, aerodynamic slip influences the flow pattern of the air. The density of the constituent material of the particles is, likewise, unimportant. Particle diameter is critical since large particles, in effect, reach out to intercept the fibres. The dimensionless parameter describing capture by interception, N_R, should, therefore, depend on particle size. To make it dimensionless, a second length must appear, and it is reasonable to identify this with the fibre diameter; and so the single fibre efficiency for capture by interception, E_R, can be expressed as,

$$E_R = E_R(N_R) \qquad (4.4)$$

where

$$N_R = \frac{d_p}{d_f}. \qquad (4.5)$$

We can make some deductions about the way that E_R varies with N_R, using simple arguments. Equation 3.9, derived from first principles, gives a stream function that varies quadratically with distance from the fibre surface, provided that the latter is small. Since ψ at any point is a measure of the amount of air flowing between the point where ψ is defined and the point where it is zero (the fibre surface) both this integrated airflow and the number of suspended particles crossing unit area in unit time will have a quadratic dependence on distance from the surface. Particles will intercept the fibre from a distance proportional to their linear dimensions, d_p in the case of spherical particles, and so the capture efficiency must approximate to a quadratic function of the particle diameter and, therefore, the interception parameter.

$$E_R \sim N_R^2 \tag{4.6}$$

Derivation of single fibre efficiency by direct interception

A more rigorous theory is as follows. Figure 4.2 shows that the particle touching the fibre at the point $\theta = \pi/2$ is on the limiting streamline. Equation 4.2 shows that E_R is given by,

$$E_R = \frac{2\psi\left(R + \frac{d_p}{2}, \frac{\pi}{2}\right)}{U\,d_f}. \tag{4.7}$$

The value of the stream function at the point $r = R + d_p/2;\ \theta = \pi/2$ depends on the particular model of filtration used. For the Kuwabara model (1), the most popular of the single fibre models,

$$E_R = \frac{1}{2Ku}\left\{2(1+N_R)\ln(1+N_R) - (1+N_R)(1-c)\right.$$

$$\left. + (1+N_R)^{-1}\left(1 - \frac{c}{2}\right) - \frac{c}{2}(1+N_R)^3\right\}. \tag{4.8}$$

The single fibre efficiency has, as required, no dependence on velocity and a quadratic dependence upon N_R as $N_R \to 0$. Values of E_R for the other single fibre models follow a similar pattern.

If the stream function takes the form given by equation 3.34, a simple approximation to the expression for E_R is, when N_R is small (2),

$$E_R \doteqdot \frac{N_R^2(1-c)}{Ku} \doteqdot \frac{N_R^2}{\zeta} \tag{4.9}$$

which has the functional form of equation 4.6. This equation is the lowest order term in a power series in N_R. A second approximation is obtained by the inclusion of an expression containing N_R in the denominator. The expression for the Kuwabara flow field, but easily generalised to other models, is (3),

$$E_R = \frac{(1-c)N_R^2}{Ku(1+N_R)^m}$$ (4.10)

where,

$$m = \frac{2}{3(1-c)}$$ (4.11)

which takes a particularly simple form when the packing fraction is small.

Numerical calculations have been carried out of the capture efficiency by interception for particles by fibres of rectangular cross section (4). In this situation the streamlines pass close to the angles and the result is an increased interception efficiency, such that E_R varies as the interception parameter is raised to a power less than 2, 1.55 in the example quoted. The increased efficiency caused by structures projecting into the flow is a subject of considerable importance, which will be studied in Chapter 8.

We have, so far, considered that for a particle to touch a fibre is sufficient to secure capture. This point will be considered in detail in Chapter 7, but it will emerge that a high velocity of impact can cause a particle to bounce. Particles captured by pure interception have the velocity of the air carrying them and, since this is zero at the fibre surface, contact by interception is a gentle process, unlikely to cause particle bounce.

Relationship between single fibre efficiency by interception and pressure drop

The calculated pressure drop in equation 3.35 and the calculated single fibre efficiency in equation 4.9 are similar in appearance, since they come about from theories that consider the same features of the flow pattern. For this reason it is only to be expected that there is some simple relationship between the two. Both the drag on a fibre and the capture efficiency by interception depend on details of the flow pattern close to the fibre; and the simple expression for the stream function in Stokes flow, given in Chapter 3, has sufficient information to indicate that (5),

$$P = \exp\left(-\frac{d_p^2 \, \Delta p}{2\pi\eta U \, d_f} \right). \tag{4.12}$$

This relationship can be shown to hold for filters that are more complex in structure than the simple model filters considered so far (6), and it is likely that it holds, to a reasonable approximation, for any filter. Since interception is a universal process, the relationship defines the lowest filtration efficiency that can be expected of a filter, though of course this minimum may be considerably exceeded in practice.

More important than this, it shows that an understanding of capture by interception does not require a detailed study of the structure of the filter. Interception depends on conditions close to the fibre surface and, though structure affects the conditions, it affects pressure drop and single fibre efficiency in roughly the same way.

Effect of aerodynamic slip on particle capture by interception

It was shown in Chapter 3 that aerodynamic slip has the effect of relaxing the boundary condition of zero tangential velocity at the fibre surface. This condition is responsible for the quadratic dependence of interception efficiency on N_R, and its relaxation introduces a linear term. An expression including slip, based on the Kuwabara flow field in the low packing fraction limit, is (7),

$$E_R = \frac{(1 + N_R)^{-1} - (1 + N_R) + 2(1 + 1.996\,Kn)(1 + N_R)\ln(1 + N_R)}{2[-0.75 - \tfrac{1}{2}\ln(c)] + 1.996\,Kn[-0.5 - \ln(c)]}. \tag{4.13}$$

The denominator of equation 4.13 is similar to that of equation 3.65, the corresponding equation for pressure drop, except that terms of order c and higher are neglected. Aerodynamic slip will occur at normal atmospheric pressure in filters made from ultrafine fibres, giving the double benefit of reduction in pressure drop and increase in filtration efficiency.

Experimental observation of particle capture by interception

Interception usually acts along with other filtration processes, but it acts more or less alone in the filters of simple respirators (8) used to give protection against dust of low toxicity. These are made from fibres of about 20 μm in diameter, and are normally operated at a face velocity of about 0.04 ms^{-1}. The results of measurements made as a simulation of their behaviour in practice, using monodisperse aerosols of a few micrometres in diameter, are shown in Fig. 4.3, which illustrates the parabolic shape of the penetration curve, characteristic of the interception process. The same basic shape of curve is also observed in situations where the geometry of the filter on a microscopic level is too complicated to be described in terms of single fibres, provided that interception is the dominant process.

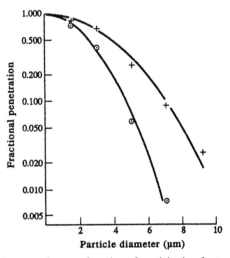

FIG. 4.3   Aerosol penetration as a function of particle size, for two filters acting by interception only (6). (© Crown Copyright; by permission of Her Majesty's Stationery Office.)

Particle capture by inertial impaction

Any convergence, divergence or curvature of streamlines involves acceleration of the air, and under such conditions a particle may not be able to follow the airflow, as shown in Fig. 4.2. What the particle does depends upon its mass or inertia (high inertia in any sense of the word implies reluctance to change) and upon the Stokes drag exerted by the air. A point particle with zero inertia will follow the streamlines and be taken past the fibre; the single fibre efficiency for inertial capture, E_I, will be zero. A point particle of infinite inertia will ignore the flow field altogether, and so E_I in this case will be unity. For the situation in between, all that a simple approach can yield is the dimensionless parameter that the efficiency depends on. In fact not only is a simple approach limited in effectiveness, the extent to which analytical approximations can be obtained is more limited for this mechanism than for any other. Much of the theory is based on computational work.

Stopping time, stopping distance and Stokes number

The dimensionless parameter describing the behaviour of a particle of finite inertia can be derived if, first, the simplified equation of motion is considered, for a spherical particle projected into a stationary medium, and experiencing Stokes drag.

$$m \frac{dV}{dt} = -3\pi\eta \, d_p V \tag{4.14}$$

The solution of this simple equation of motion gives the particle's stopping time, τ_s, the time required for its velocity to drop by a factor of e, and its stopping distance, d_s, the distance it travels before coming to rest:

$$\tau_s = \frac{d_p^2 \rho}{18\eta} \tag{4.15}$$

and

$$d_s = \frac{V d_p^2 \rho}{18\eta}. \tag{4.16}$$

$d_p\sqrt{\rho}$ is the aerodynamic diameter of a spherical particle, and the same quantity for a particle of any shape can be obtained from experimental measurement of τ_s or d_s. Particles with large aerodynamic diameters will be those least able to follow streamlines.

The stopping time and the stopping distance are still relevant parameters in the situation in which the air into which they are projected is, itself, moving. The parameter describing the behaviour of a particle suspended in an airstream of velocity U is given by replacing V by U in equation 4.16, and the dimensionless parameter describing capture by inertial impaction is the quotient of this and a characteristic length describing the filter. Just as in filtration by interception, this characteristic length is the fibre diameter. The result is the Stokes number St.

$$St = \frac{d_p^2 \rho U}{18\eta \, d_f} \tag{4.17}$$

Rigorous theory of inertial effects

Rigorous study of inertial impaction requires the solution of the equations of motion of the particle travelling in the air and subject to drag when the response of the air is fully taken into account (9):

$$m \frac{dV}{dt} = -3\pi\eta \, d_p V - \frac{\pi}{12} \rho_g \, d_p^3 \frac{dV}{dt}$$

$$- \tfrac{3}{2} d_p^2 \sqrt{\pi\eta\rho_g} \int_o^t \frac{dV}{ds} \frac{ds}{\sqrt{t-s}} \tag{4.18}$$

where ρ_g is the density of the air. Equation 4.18 can easily be generalised to include an external force, but as it stands, the first term on the right hand side is simply the Stokes drag on the particle, corresponding to that in equation 4.14. The second term arises because not only does the particle accelerate, but so does the air surrounding it when it is displaced by the particle's motion. This increases the effective mass of the particle by one half of the mass of air displaced, but this is usually negligible because the density of the air is much lower than that of the particle (it is important in studies of particles moving in liquids of comparable density, and especially important for bubbles moving in liquids of much higher density). The third term accounts for the effects of past accelerations on the particle, which are weighted according to the reciprocal of the square root of the elapsed time. It is associated with diffusion of vorticity from the particle (10). The solution of the exact equation of motion of particles moving in air can be extremely difficult, but in practice both the added mass term and the history term may be neglected. The simplified equation, for a particle moving in air which is itself moving, is,

$$\frac{dV}{dt} = \frac{18\eta}{d_p^2 \rho} (U - V). \tag{4.19}$$

Even equation 4.19 has no general analytical solution because of the complexity of U, the flow field, and so it must be solved numerically. Fortunately, this is not difficult; and the most popular procedure is the Runge-Kutta method of stepwise solution (11).

If the particle trajectory, calculated in this way, touches the fibre surface, capture is assumed to occur, though in fact this involves two approximations. The first is that the hydrodynamic interaction between the particle and the fibre is neglected. As the particle approaches closely, the air between it and the fibre has to be rapidly displaced, an effect resembling the added mass effect, but like it, a complication that can normally be neglected in the filtration of aerosols (12, 13). The second approximation is that the particles are assumed not to bounce. Situations in which this assumption is not valid will be described in Chapter 7.

Inertial impaction at low Stokes number

Calculations show that in the absence of interception, particles with a Stokes number smaller than some critical value, which depends on the flow field, are captured at a negligible rate. A mathematical approach (14) predicts that capture efficiency remains finite, though its value is extremely small.

In practice, as described in the section above, interception can never be completely neglected. When the interception efficiency is finite and the Stokes number is small, as shown in Fig. 4.4, it is possible to calculate the single-fibre efficiency on the assumption that inertia causes a small perturbation in the particle's trajectory. The velocity of the particle can be equated to that of the air plus a small additional component, V', which is proportional to the Stokes number (15).

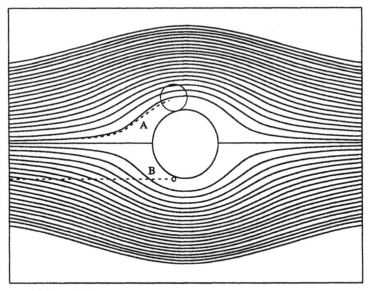

FIG. 4.4   Schematic illustration of particle trajectories under perturbation approximation: A, particle captured by interception with inertial impaction as a small perturbation; B, particle of high Stokes number perturbed very little by the airflow.

$$V = U + St V' \tag{4.20}$$

This makes it possible to derive an expression for the single fibre efficiency for capture of the particles, by integrating along the fluid streamlines.

$$E_{IR} = E_R + (2\zeta)^{-2} JSt \tag{4.21}$$

The constant J, which is obtained by numerical integration, can be fitted by the following analytical expression, provided that $0.01 < N_R < 0.4$ and $0.0035 < c < 0.111$, for the Kuwabara model.

$$J = (29.6 - 28\ c^{0.62}) N_R - 27.5\ N_R^{2.8} \tag{4.22}$$

Inertial impaction at high Stokes number

When the Stokes number is high, the trajectory of a particle is almost a straight line, as illustrated in Fig. 4.4, and the velocity of the particle is close to its initial velocity, usually assumed to be the velocity of the air at large distances from the fibre (16). If the drag forces acting on it during its real trajectory are approximated by those acting on a particle moving in a straight line, the equation of motion, 4.19, can be solved easily. This approach, like that adopted with particles of small Stokes number, is first order perturbation theory, but the working is considerably easier because the approximate trajectory is much simpler. The result is,

$$E_1 = 1 - \frac{\mu}{St} \tag{4.23}$$

where the constant μ depends on the flow field. For the Kuwabara field with $c = 0.05$ it has the value 0.805.

Calculation of inertial impaction efficiency over a range of Stokes numbers

A number of workers (17, 18, 19, 20, 21, 22, 23) have carried out calculations of inertial impaction, the general results of which are consistent with the above approximations, in that the single fibre efficiency increases slowly with the Stokes number when the latter is small, and that it approaches unity asymptotically when the latter is large. In between there is a region over which increase is rapid, much more rapid than that occurring as a result of capture by interception. (The high size-selectivity of the process in general makes it useful in dust sampling involving inertial separators.) The capture efficiency by inertial impaction increases with increasing packing fraction and with increasing Reynolds number.

Curves 1 and 2 (17) in Fig. 4.5 show all of these aspects of behaviour (except for Reynolds number dependence) in calculations based on the Kuwabara model. The curve for single fibre efficiency due to interception, drawn on the same graph, would be a straight line of unit gradient since single fibre efficiency by this process depends on d_p^2, just as does the Stokes Number. The results of calculations based on the small Stokes number approximation (15), reported by another observer (2) are also shown in Fig. 4.5.

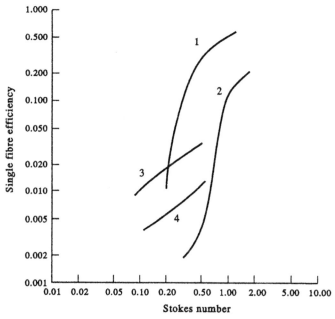

FIG. 4.5 Calculated values of single fibre efficiency by inertial impaction: (1) reference 17, $c = 0.11$; (2) reference 17, $c = 0.01$; (3) reference 15, $c = 0.1$, $N_R = 0.05$; (4) reference 15, $c = 0.01$, $N_R = 0.05$. (By permission of Pergamon Press, Headington Hill Hall, Oxford OX3 0BW, UK, and Academic Press, London.)

Formulae for single fibre efficiency fitted to the results of calculation and experiment

Fitting analytical expressions to data obtained for filtration efficiency by inertial impaction is fundamentally more difficult than it is for other filtration processes. Simple theoretical models exist for all other processes, and these lead to expressions, such as equation 4.9, relating single fibre efficiency to the appropriate dimensionless parameter. An expression with the same functional form but variable coefficients can then be fitted either to experimental data or to the results of computation.

Alternatively a simple power relationship may be assumed between single fibre efficiency and the dimensionless parameter. In the case of inertial impaction, theory gives an analytical description only in limiting situations, and description of the behaviour for moderate values of St requires curve-fitting. An example of the results of this procedure is the following equation (24) fitted to calculations based on an isolated fibre model (25).

$$E_1 = \frac{St^3}{St^3 + 0.77St^2 + 0.22} \tag{4.24}$$

This expression has the characteristic shape, and the same analytical form as equation 4.23 in the high St limit; and both are plotted in Fig. 4.6. The expression has formed the basis of curve-fitting to experimental results, including the effect of both packing fraction and Reynolds number (18):

$$E_1 = \frac{St^3 e^3}{St^3 e^3 + 0.77\left(1 + \dfrac{K_3}{Re^{1/2}} + \dfrac{K_4}{Re}\right)St^2 e^2 + 0.58} \tag{4.25}$$

where $e(c)$ is a function of the packing fraction, given by,

$$e(c) = 1 + K_1 c + K_2 c^2. \tag{4.26}$$

In equation 4.26, the parameters K_1–K_4 were fitted by least squares to experimental data on real and model filters separately, and two sets, listed in Table 4.1, result. The fact that these differ widely for the two structures indicates the difficulty in obtaining a generalised expression for particle capture by inertial impaction. Curves derived from equations 4.24 and 4.25 are plotted on Fig. 4.6 for various values of packing fraction and Reynolds number.

Experimental measurements on model and real filters

In an exact calculation, the equation of motion must be solved as a particle passes through the flow field, and so the motion will be sensitive to details in the flow pattern. Unfortunately the structure of real filters is highly complex, whereas that of the systems chosen to

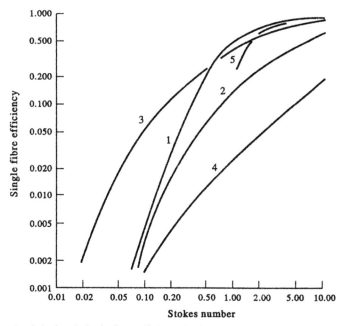

FIG. 4.6 Calculated single fibre efficiency by inertial impaction: (1) according to reference 24; (2) reference 18 (model filters), $c = 0.01$, $Re = 10$; (3) reference 18, $c = 0.1$ $Re = 10$; (4) $c = 0.01$, $Re = 1.0$; (5) reference 16.

TABLE 4.1
Fitted parameters to equation 4.25, taken from reference 18

Filter type	K_1	K_2	K_3	K_4
Model filter	37	91	12	60
Real filter	4	2250	4	65

simulate them is usually rather over-simplified, and this means that a direct comparison between theory and experiment for a real filter is not easy. In studies of inertial impaction, measurements on model filters form a useful bridge between theory and experiment.

Measurements made with woven mesh filters (23) are plotted on Fig. 4.7, the filters having packing fractions between 0.006 and 0.025, and being operated at Reynolds numbers between 30 and 80. On the same figure are plotted results on model filters (26), constructed from stacks of sheets of straight parallel and equally spaced fibres of about 9 μm in diameter, spaced sufficiently far apart to ensure that the effects of the flow pattern due to each layer would simply be additive, and that no interference would take place. These were exposed to monodisperse liquid aerosols (chosen to ensure that capture was 100% effective and that no bouncing took place) at low Reynolds number (the results were found to compare well with theoretical calculations based on the complex variable model (27)).

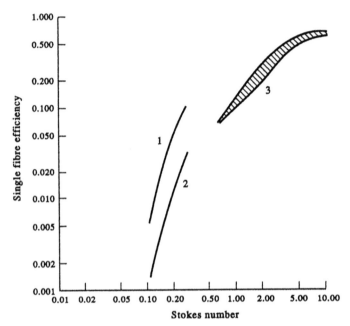

F‍IG. 4.7   Experimentally measured single fibre efficiency by inertial impaction: (1) reference 23, $c=0.025$, $Re=30\text{–}80$; (2) reference 23, $c=0.006$, $Re=30\text{–}80$; (3) reference 26 envelope curves for single meshes, $\pi R/2b=0.223$, $Re\to0$. (By permission of the Society of Chemical Engineers of Japan, and Pergamon Press, Headington Hill Hall, Oxford OX3 0BW, UK.)

Further experimental data are plotted in Figs 4.8 to 4.10, several figures being used, in order to prevent the overlap of plotted data. Two sets of data are plotted on each figure, though these are not necessarily related, since the choices have been made for the sake of clarity. However, taken together, the figures should give a picture of the general pattern and the measure of variation among results obtained on filtration by inertial impaction.

Figure 4.8 shows two sets of data for parallel wire model filters with packing fractions of 0.01 and 0.11, operated at Reynolds numbers of 1–5 (20). These are plotted along with data for woven mesh filters operated at low Stokes number. The parameter specifying the structure of a single layer of mesh, and corresponding most closely to the packing fraction of a filter composed of many such layers is $\pi R/2b$, where R is the fibre radius and $2b$ the interfibre spacing; and this varies between

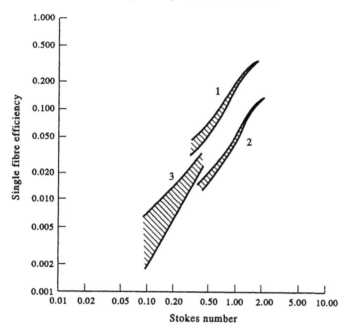

FIG. 4.8 Experimentally measured single fibre efficiency by inertial impaction: (1) reference 20, $c = 0.11$; (2) reference 20, $c = 0.01$; (3) reference 28 single mesh results. (By permission of Pergamon Press, Headington Hill Hall, Oxford OX3 0BW, UK, and the American Institute of Chemical Engineers, © A.I.Ch.E. all rights reserved.)

0.6 and 0.8 for the meshes in question, which were operated at Reynolds numbers between 1 and 4 (28).

Figure 4.9 has data on woven mesh filters with the above grid-specifying parameter of 0.7 and Reynolds numbers of 1–45 (29), and these are plotted along with experimental data obtained with polyester fibre filters (30).

Finally, in Fig. 4.10 two sets of data for woven mesh filters are plotted together. In one case (31) the packing fraction varies between 0.20 and 0.25, and the Reynolds number between 1 and 10. In the other (32), the grid-specifying parameter is 1.08 and the Reynolds number 0.5–10.0.

The data largely speak for themselves, though at small Stokes number in Fig. 4.10 the single fibre efficiency increases as *St* decreases. The same phenomenon, observed by other workers (20), is attributed to the effects of electric charge, which should become more apparent at smaller particle sizes. This aspect of filtration will be treated at length in Chapter 6.

Other data (32) show a weak variation of single fibre efficiency with Stokes number; and in a further study of the behaviour of particles with a Stokes number of up to and exceeding 100 (33, 34), the single fibre

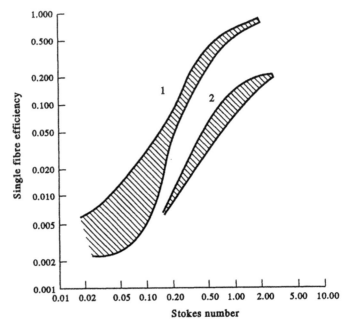

FIG. 4.9 Experimentally measured single fibre efficiency by inertial impaction: (1) envelope of data from reference 29; (2) envelope of data from reference 30. (© 1982, Elsevier Science Publishing Co. Inc.)

capture efficiency decreased as Stokes number increased, because of particle bounce, as briefly mentioned above.

Inertial impaction is probably the most poorly understood filtration process. No analytical approximations exist except in the limits of low and of high Stokes number. For the range of values of particular interest, where the theoretical single fibre efficiency varies rapidly with the Stokes number, only semi-empirical formulae exist, and even these are limited in application. It seems likely that, since the single fibre efficiency depends on so many variables including, in real filters, the elusive variable of structure, no single expression can give a good general description of inertial impaction. Many observations show less variation in filtration efficiency with Stokes number than calculations would suggest. We might seek to explain this in terms of the variation of conditions throughout the filter; but it is clear that a simple single-fibre approach is insufficient for studies of filtration by inertial impaction.

Capture of particles by gravity

Aerosol particles in still air tend to settle out under the influence of gravity, and this process will also occur when the air suspending them is

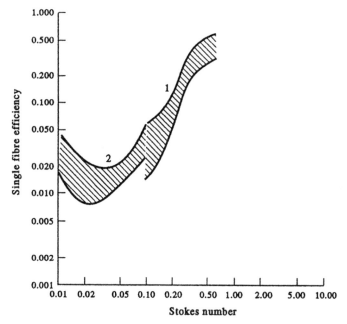

FIG. 4.10 Experimentally measured single fibre efficiency by inertial impaction: (1) reference 31, $c = 0.20$–0.25, $Re = 10$; (2) reference 32, $\pi R/2b = 1.08$, $Re = 0.5$–1.0. (By permission of Academic Press, London, and Pergamon Press, Headington Hill Hall, Oxford OX3 0BW, UK.)

passed through a filter. If, as a result of this process, a particle contacts a fibre it will be captured—by gravitational deposition. The effect of gravity during filtration will depend on the direction of airflow, with the result that gravitational settling may either augment or diminish the transport of the particles towards the fibres.

Simplified approach to gravitational capture

Again, our first approach to the problem will be by way of simple concepts. The settling velocity of a particle under the influence of gravity is,

$$V_{G} = \frac{d_{p}^{2}\rho g}{18\eta} \tag{4.27}$$

where g is the acceleration due to gravity. Whereas the settling velocity will in most cases tend to augment capture, the other principal velocity component, the convective velocity of the air through the fibres, will tend to carry particles past the fibres. The general effectiveness of gravitational capture will depend on the relative sizes of the two, and the dimensionless quantity describing capture will be the quotient,

$$N_G = \frac{d_p^2 \rho g}{18 \eta U}. \tag{4.28}$$

Equation 4.28 applies strictly to spherical particles, but substitution of the settling velocity by way of equation 4.27 results in a dimensionless parameter appropriate for particles of all shapes.

If gravity acts alone, doubling the convective velocity would give the force only half the time to act, and intuitively one is led to think that this should reduce the capture efficiency by a factor of two, suggesting that the single fibre efficiency is directly proportional to N_G when gravity acts alone.

$$E_G \sim N_G \tag{4.29}$$

Experimental observation of gravitational capture

Experimental measurements on porous foam with pore sizes of 20–30 ppi (35, see also 36) using monodisperse aerosols gave results for layer efficiency that are plotted against filtration velocity in Fig. 4.11. The layer efficiency and the filtration velocity are plotted on logarithmic scales, which means that a power law relationship would result in a straight line with a gradient equal to the power. The results show that the layer efficiency and, therefore, the single fibre efficiency, varies as U^{-1}, which is exactly the relationship given by equations 4.28 and 4.29. A plot of the same data against particle diameter reveals a quadratic dependence, confirming the conclusion.

Theory of gravitational capture

The theory of gravitational capture can be worked through analytically (37, 38, 39), because gravitational forces are solenoidal. As mentioned in Chapter 3, a solenoidal field is one in which the divergence is zero, a property also satisfied by the airflow velocity field through the filter.

$$\nabla \cdot \underline{U} = 0 \tag{4.30}$$

This means that the velocity of the particles under the influence of gravity could be described by a stream function, satisfying equations 3.2 or 3.3. The streamlines for the motion of particles under gravity alone will simply be vertical lines, describing the trajectories of settling

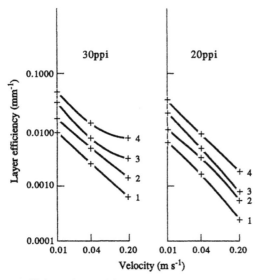

Fig. 4.11 Layer efficiency for particle capture by gravitational settling. Results on two grades of porous foam, illustrating the dependence of E_G on velocity and therefore, dimensionless parameter N_G: (1) 3 μm particles; (2) 5 μm particles; (3) 7 μm particles; (4) 9 μm particles (35). (© Crown Copyright; by permission of Her Majesty's Stationery Office.)

particles; and the stream function for airflow and gravity together can be obtained by adding the gravitational stream function and the fluid flow stream function, as shown in Fig. 4.12. Expressions for the particle trajectories will, therefore, exist in an analytical form, making stepwise trajectory calculations unnecessary; and this means the single fibre efficiency can be written down from the values of the stream function for the limiting trajectories. The result is,

$$E_G = N_G. \qquad (4.31)$$

Capture of particles by diffusional deposition

The capture processes described so far are more efficient for large particles. An important capture process that is more efficient for small particles comes about by the diffusional motion of the particles. Thermal energy in a gas at equilibrium is distributed among the molecules of the gas so that every degree of freedom has associated with it an energy $\frac{1}{2}k_B T$, where k_B is Boltzmann's constant and T is the absolute temperature (40). Particles suspended in the gas will rapidly come into thermal equilibrium with it, so that they also receive their share of thermal energy. A constant exchange of energy between a particle and gas molecules results in the familiar microscopic motion

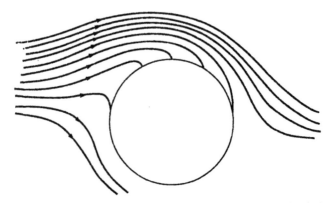

Fig. 4.12 Calculated particle trajectories for gravitational deposition in horizontal flow.

known as Brownian motion. Since equipartition of energy means that the average kinetic energy of a particle is a constant, independent of its size, the average velocity of small particles as a result of this motion will be greater than that of large particles.

Diffusive motion of particles

The motion of particles can be quantified by a coefficient of diffusion, D, which is similar in nature to the diffusion coefficient for molecules but much smaller. D is defined below (9), in terms of the average displacement of a particle from its original position in any direction, the x-direction say, in a time, t,

$$\overline{x^2} = 2Dt. \tag{4.32}$$

The coefficient of diffusion of the particles is related to their mobility, μ, by the Einstein equation,

$$D = \mu k_B T \tag{4.33}$$

where,

$$\mu = \frac{Cn}{3\pi\eta\, d_p}. \tag{4.34}$$

Cn is the Cunningham correction factor (41, 42) necessary because of aerodynamic slip at the particle surface, and normally defined by,

$$Cn = 1 + \frac{2A\lambda}{d_p} + \frac{2Q\lambda}{d_p} \exp - \frac{Bd_p}{2\lambda} \qquad (4.35)$$

where $A = 1.246$, $Q = 0.42$, $B = 0.87$ and λ, the mean free path of molecules at NTP, is 0.065 μm. Aerodynamic slip increases the mobility of small particles just as it reduces the drag on fine fibres. Since, however, a filter is often set to capture particles with much smaller diameters than those of its own fibres, aerodynamic slip is more significant in particle dynamics than in flow field specifications. The diffusion coefficient of spherical particles of a range of diameters, calculated according to equation 4.33, is shown in Fig. 4.13. Alternative (43, 44), slightly different values for the constants in equation 4.35, $A = 1.142$, $Q = 0.558$, $B = 0.999$, have been quoted, as a result of measurements made with solid aerosols, more likely to suffer specular reflection of air molecules (see section on slip in Chapter 3).

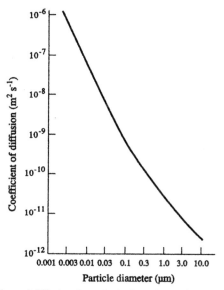

FIG. 4.13 Coefficient of diffusion for spherical particles of various sizes, calculated from equations 4.33–4.35.

The capture of particles by a fibre, as a result of this process, will depend on the relative magnitude of the diffusional motion and the convective motion of the air past the fibre; and the dimensionless parameter relating these is the Peclet number, Pe (45),

$$Pe = \frac{2UR}{D}. \qquad (4.36)$$

The way that the Peclet number is defined means that particle capture efficiency by diffusion will decrease as the Peclet number increases.

Simplified calculation of single fibre efficiency

The exact calculation of the single fibre efficiency requires the solution of a complicated transport equation, but first a very simple approximation will be made, in order to obtain an understanding of the place of the various physical quantities in the formulation of the capture efficiency. It frequently happens that the salient features of a process emerge from a very crude approach, and although what follows below lacks rigour, it should help in the understanding of more exact derivations.

Analytical theories consider only the situation where the single fibre efficiency is small, and where the particles that are captured are those which, even in the absence of diffusional behaviour, would pass close to the fibre. The approximation to the stream function, close to the fibre (equation 3.34)

$$\psi = \frac{U \, \delta^2 \sin \theta}{R \zeta} \tag{4.37}$$

is taken as a starting point, and the coordinates δ and θ are as shown in Fig. 4.14. Two rather crude approximations must now be made, the first being that the angular dependence of ψ can be averaged, meaning that the air velocity in the θ direction is given by,

$$U_\theta = \frac{2U \, \delta}{R \zeta}. \tag{4.38}$$

Equation 4.38 implies effectively zero radial velocity, which is rather like approximating the fibre surface by a planar surface undergoing Poiseiulle flow. The second approximation is that the random nature of the diffusional motion can be replaced by ordered motion, with the particle at the mean position. The mean displacement follows from equation 4.32.

$$\delta = \sqrt{2Dt} \tag{4.39}$$

With these approximations it is a simple matter to solve the trajectory equation of the particle in reverse, starting at the rear stagnation point. The velocity of the particle in the θ-direction equals

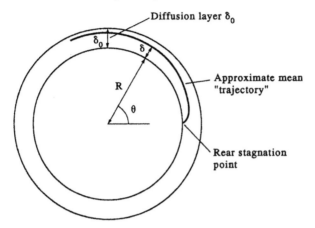

FIG. 4.14 Coordinates, approximate trajectory and diffusion layer, calculated from elementary theory of diffusional capture.

U_θ in equation 4.38 and that in the δ-direction is obtained by differentiating equation 4.39.

$$V_\delta = \frac{D}{\delta} \tag{4.40}$$

Combining the two equations gives,

$$\frac{d\delta}{d\theta} = \frac{R\dfrac{d\delta}{dt}}{\dfrac{d\theta}{dt}} = \frac{RV_\delta}{U_\theta} = \frac{D\zeta R^2}{2U\,\delta^2} \tag{4.41}$$

which can be simply integrated to give,

$$\delta_0 = \left(\frac{3D\zeta R^2 \pi}{2U}\right)^{1/3} \tag{4.42}$$

in which the integration has been performed between the limits of $\theta=0$ and π. The value of ψ at the point where the particle originates is given by the equation for the stream function; and the capture efficiency is related to ψ by equation 4.2. The single fibre capture efficiency in this simple model is, therefore, given by,

$$E_D = 4.46\ \zeta^{-1/3} Pe^{-2/3}. \tag{4.43}$$

The result is neither rigorous nor exact, but the functional forms in which the hydrodynamic factor and the Peclet number appear in the expression for single fibre efficiency are clearly given, and are in fact the same as those predicted by more sophisticated theory. The values of the powers appearing in the equation result from the Stokes flow form of the velocity profile near to the fibre surface; and the numerical coefficient differs, but not vastly, from that resulting from rigorous calculation. Moreover, it is also clear that δ_0 in equation 4.42 is related to the thickness of the layer round the fibre from which particles are effectively removed by diffusional deposition. The thickness of this layer, the diffusion layer, is illustrated in Fig. 4.14.

Rigorous theory of diffusional capture

The exact equation describing the evolution of the average density of particles under the influence of diffusion and convection is the Fokker-Planck equation (9):

$$DV^2N - \nabla \cdot (\underline{U}N) = \frac{\partial N}{\partial t}. \tag{4.44}$$

In the situation of diffusional capture of particles by a fibre from an aerosol flowing through a filter, steady state conditions can be assumed to apply, and so the right hand side of the equation can be set equal to zero; and in the case of flow around a fibre, the cylindrical symmetry enables the equation to be simplified,

$$U_r \frac{\partial N}{\partial r} + \frac{U_\theta}{r} \frac{\partial N}{\partial \theta} = D \left(\frac{\partial^2 N}{\partial r^2} + \frac{1}{r} \frac{\partial N}{\partial r} + \frac{1}{r^2} \frac{\partial^2 N}{\partial \theta^2} \right). \tag{4.45}$$

Reduction of this equation to dimensionless parameters causes the Peclet number to emerge. Furthermore, it is usual practice to neglect the third term on the right hand side of the equation; which is reasonable if the diffusion layer is small, and which is equivalent to simplifications made in studies of diffusional transport in systems of simpler geometry (46).

As it stands, even the simplified equation has no simple solution, but it can be integrated (47) by change of variable, provided that the second term on the right hand side of the equation is neglected as well, giving,

$$E_D = 2.32 \, \zeta^{-1/3} Pe^{-2/3}. \tag{4.46}$$

Further approximations by a number of authors have been reviewed (48, 49), all giving basically similar equations, and there is a general consensus around the expression,

$$E_D = 2.9 \, \zeta^{-1/3} Pe^{-2/3}. \tag{4.47}$$

A more detailed solution, involving numerical integration, takes account of the second term on the right hand side of equation 4.45. Close to the fibre surface, this term is more significant than the angular derivative, but less significant than the first term (50).

$$E_D = 2.9 \, \zeta^{-1/3} Pe^{-2/3} + 0.62 Pe^{-1} \tag{4.48}$$

The thickness of the diffusion layer can be calculated by this approach, the value being smaller than δ_0 in equation 4.42 by a factor of 0.75.

$$\delta_0 = \left(\frac{4\zeta}{Pe} \right)^{1/3} R \tag{4.49}$$

Numerical calculations of particle trajectories (51) have resulted in an expression for E_D as a multiple of Pe raised to the power 0.9; and theoretical calculations of the effect of fibre shape on single fibre efficiency (52) predicted that fibres with a circular cross-section were marginally more efficient collectors of particles than were fibres with elliptical cross-section, irrespective of the orientation of the axis of the ellipse with respect to the airflow. Other calculations, based on the assumption of rectangular fibre cross-section (4), predicted that the single fibre efficiency, which has the same dependence on Peclet number as it does for cylindrical fibres, is greater for fibres orientated with their long edge parallel to the flow than for those with the opposite orientation.

A further theoretical study, of diffusional deposition on fibres that are not perpendicular to the flow (53), predicted that the capture efficiency becomes progressively lower as the inclination of the fibres deviates from perpendicular. In the extreme, of fibres parallel to the flow, single fibre efficiency ceases to be a valid concept.

Experimental observation of diffusional capture

Equation 4.47 has been verified (50) by experiments on model filters of the staggered geometry, using aerosols of sodium chloride and Di-octyl Sebacate (DOS). Very good agreement between theory and experiment was found for values of Pe greater than about 2, which is a less rigorous criterion than that strictly required by the theory. In particular, the functional dependence of E_D on Pe was very clearly indicated by the experimental results. Polydisperse aerosols were used in the experiments, detected by diffusion batteries, and so particles were selected according to the physical parameters on which filtration efficiency depended (see Chapter 9).

Experiments were also carried out with fan model filters, the structure of which is closer to that of real filters than the staggered arrays of parallel fibres. The observed functional dependence of E_D on *Pe* was the same but the numerical coefficient was different.

$$E_D = 2.7 Pe^{-2/3} \qquad\qquad (4.50)$$

Real filters were found to be rather less efficient than fan model filters, a fact accountable to their less than perfect structure. A correlation between efficiency and pressure drop would be expected, because of the presence of the hydrodynamic factor in the expression for single fibre efficiency; and the experimental results showed the single fibre efficiency to be proportional to the pressure drop.

Theory and experiment were compared (30) for real filters made from polyester fibres, and good agreement was found between the experimental results and the predictions of equation 4.47, though theory tended to over-estimate efficiency, especially at small values of *Pe*.

Measurements of the efficiency of stacks of woven mesh (54) with packing fractions between 0.02 and 0.35 agreed very closely with equation 4.50, when a power law in *Pe* was fitted. The exponent was −2/3 and the coefficient varied between 2.69 and 2.71.

Equation 4.50 was found to be valid for *Re* < 1 (55) but efficiency was found to increase at higher values of Reynolds number, being approximately doubled at a Reynolds number of 10. In other work (56) theoretical predictions and experimental results were found to agree when *Pe* was greater than about 0.1, provided that the Reynolds number was less than 1 and the packing fraction between about 0.017 and 0.35; and in fact these criteria hold for a wide range of filters used in practice.

At very low Peclet numbers, the calculated width of the diffusion layer would exceed half the inter-fibre spacing, and overlap of diffusion layers would mean that the approximations used in the derivation of the expressions for E_D would cease to be valid. Nevertheless, equation 4.47 accurately described experimental measurements in this situation. This observation is just one of many examples of a physical law or expression that holds well outside the region of parameters where it can be rigorously justified.

Effect of aerodynamic slip on diffusional capture

Calculation of the effect of aerodynamic slip on single fibre efficiency by diffusional capture (57) starts with equation 4.45 with the second and third terms on the right hand side neglected. The boundary condition of zero tangential velocity at the fibre surface is replaced by one which gives a slip velocity proportional to the tangential stress as in equation 3.19. A considerable amount of mathematical working is required, but the result is that,

$$E_{\mathrm{D}} = 2.27 \; \zeta^{-1/3} Pe^{-2/3}\left(1 + 0.62 \frac{KnPe^{1/3}}{\zeta^{1/3}}\right). \tag{4.51}$$

In the absence of slip this term does not quite revert to the accepted form, but this is attributed to approximations involved during the numerical work. An expression with a similar functional form but in which the numerical coefficients are 2.7 and 0.55 respectively has also been derived (58).

Experimental measurements of the effect of slip have shown that it tends to increase the filtration efficiency, as would be expected (59). These results followed the basic form given by equation 4.48 but with ζ modified according to the simple dependence on pressure given by equation 3.65.

Capture of non-spherical particles by fibrous filters

Both theory and experiment have tended to concentrate on spherical particles, for these are the easiest to describe and to produce in controlled ways; but the same restriction cannot be placed on the use that filters are put to. Aerosols, particularly agglomerated aerosols, may exist in a range of elaborate shapes, the simplest and most important non-spherical particles being fibres. The health hazards of asbestos and to some extent other fibrous aerosols are well-known, and their removal by filtration is widely practised.

Dynamics of fibrous aerosols

Fibres or linear agglomerates of simple particles can be captured by mechanisms that are similar or analogous to those that act in the filtration of spherical particles, but the parameters describing their behaviour are more complicated. The interception parameter and Stokes number of spherical particles are scalar quantities; a single number is sufficient to quantify their behaviour. When applied to non-spherical particles, the aerodynamic diameter becomes a tensor, for the response of an irregularly shaped particle to a force in one direction is not necessarily the same as its response to the same force in another direction. However, the relatively high symmetry of fibres and linear agglomerates means that two parameters, corresponding to the directions parallel to and perpendicular to the fibre axis, are sufficient to describe their behaviour. The parameters appropriate to interception, inertial impaction and diffusional deposition will be considered in turn.

Interception and fibre aspect ratio

The interception parameter of a sphere is proportional to its diameter. The two corresponding quantities for a fibre are proportional to its length and to the diameter of its cross section, d_e. The quotients of the lengths or the dimensionless parameters are equal to the aspect ratio of the fibre, β, and this may be considerable. Figure 4.15 illustrates that the probability of a fibre being captured by interception as it passes through the filter depends critically upon its orientation and, therefore, on which of the two parameters above describes its behaviour. The important property of orientation in shear flow will be described in greater detail later.

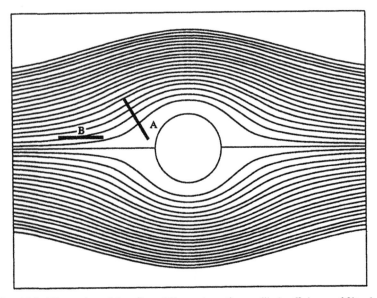

FIG. 4.15   Illustration of the effect of fibre orientation on likely efficiency of filter by interception. Fibres A and B are approximately the same length, but A is much more likely to be captured.

Inertial impaction and aerodynamic diameter

Aerodynamic diameter is closely related to mobility which, for a fibre, is given by (9),

$$\mu = \frac{1}{3\pi\eta \, d_e K}. \tag{4.52}$$

Analytical expressions for the shape factor, K, do not exist for fibres, but they have been worked out for a simple approximation to that shape, a prolate ellipsoid of revolution. In this case, d_e in equation 4.52 is replaced by the equatorial diameter, the shape factor in the direction parallel to the major axis being given by,

$$K_{\parallel} = \frac{4(\beta^2 - 1)}{3\left[\dfrac{2\beta^2 - 1}{\sqrt{\beta^2 - 1}} \ln(\beta + \sqrt{\beta^2 - 1}) - \beta\right]} \tag{4.53}$$

and in the direction perpendicular to the axis,

$$K_{\perp} = \frac{8(\beta^2 - 1)}{3\left[\dfrac{2\beta^2 - 3}{\sqrt{\beta^2 - 1}} \ln(\beta + \sqrt{\beta^2 - 1}) + \beta\right]}. \tag{4.54}$$

Figure 4.16 illustrates the variation of these with the aspect ratio of the fibre. The data in the figure show that the shape factor, and hence the mobility and the aerodynamic diameter, is much less dependent on direction than are estimates of geometric size.

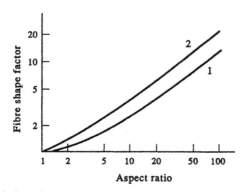

FIG. 4.16   Variation of shape factor of a prolate spheroid with aspect ratio: 1 shape factor $K_{\parallel}$; shape factor $K_{\perp}$.

Diffusive motion in translation and rotation

The diffusion of fibres is particularly complicated because it comprises both rotational and translational components. The two translational components of the coefficient of diffusion are given by substituting equations 4.52 to 4.54 into equation 4.33. Diffusion, therefore, has a directional component, the magnitude of which is given by the data in Fig. 4.16, but it may be suitable to average the two components with a weighting factor, depending on the average orientation of the fibre. Within this approximation, the theory of diffusional capture follows in much the same way as for spherical particles.

The angular component of Brownian motion is given by an expression similar to equation 4.32 (9).

$$\overline{\theta^2} = 2k_B T \mu_\omega t \qquad (4.55)$$

The angular mobility, μ_ω, is the factor relating an applied couple acting on the fibre to the angular velocity that it causes. Again, analytical expressions exist for prolate spheroids though not for fibres proper. The angular mobility of a sphere, a prolate spheroid with unit aspect ratio, is,

$$\mu_\omega = \frac{1}{\pi \eta \, d_p^3} \qquad (4.56)$$

and the same parameter for a prolate spheroid with non-unit aspect ratio is (60),

$$\mu_\omega = \frac{3 \left[\dfrac{2\beta^2 - 1}{\sqrt{\beta^2 - 1}} \ln(\beta + \sqrt{\beta^2 - 1}) - \beta \right]}{2\pi \eta \, d_e^3 (\beta^4 - 1)}. \qquad (4.57)$$

The presence of β at the fourth power in the denominator indicates that the angular mobility falls off rapidly with increasing aspect ratio. The effect of angular diffusion on capture efficiency will be through the way that it affects fibre alignment. It will tend to act against forces that cause alignment and may, as a result, tend to reduce filtration efficiency in the case of processes that align fibres in such a way as to increase their capture rate, and to increase it in the contrary situation.

Alignment of fibres in shear flow

Shear flow will exert a couple on any particle suspended in the flowing fluid; and the magnitude of the couple will depend on the velocity gradient and the size of the particle (61). The theory is precise for the situation of a prolate spheroid in uniform shear flow (9), and in this situation the angular velocity of a spheroid is given by the expression:

$$\omega = \frac{d\theta}{dt} = \Gamma \frac{\cos^2\theta + \beta^2 \sin^2\theta}{1 + \beta^2} \tag{4.58}$$

where Γ is the velocity gradient at the position of the fibre. The angular velocity will be greatest when the fibre is perpendicular to the flow direction, and so the spheroid will spend most of its time aligned parallel to the flow, but it will flip from time to time, at a frequency that varies inversely with its length. This behaviour has been observed experimentally (62), the period of rotation of cylindrical fibres being observed to be about two thirds of the calculated period of prolate spheroids of the same aspect ratio.

Measurements of filtration efficiency against fibrous aerosols

Experimental measurements have the special difficulty that fibres have to be characterised by length and diameter; and the methods for the production of monodisperse fibres, detailed in Chapter 9, are not easy to implement.

Measurements of the penetration of fibres with an aspect ratio of 25 were made using a fine sieve as a model filter (63). The results were compared with measurements made with isometric particles, and the conclusions were that the fibres penetrated to the same extent as particles with a diameter 0.285 times the fibre length. This behaviour would be anticipated if the fibres passed through at random orientations, and were captured by pure interception, and so it indicated a lack of alignment in the conditions of the experiment.

Other measurements of the penetration of linear agglomerates through model filters made from grids (64) illustrated, first of all, that the aerosols produced were relatively monodisperse, since the results showed an exponential variation of penetration with depth. The various mechanisms operating would have the same velocity dependence as those for spherical particles. In particular, if Stokes flow conditions applied both to the airflow pattern through the filter and to the rotation of the fibres in that flow, increasing the filtration velocity would cause the fibres to describe exactly the same trajectory, but at a faster rate; and so capture by interception would be velocity-independent, just as it is with spherical particles. The authors observed that the penetration increased with velocity.

Measurements on real filters are difficult to tie in with theory, but some measurements have been made, using polydisperse fibrous aerosols and filters of incompletely specified structure, principally for the purpose of estimating protection of respirators against asbestos (65, 66, 67). The measurements were compared with isometric aerosol particle penetration; and a pattern seemed to emerge, that the fibres may be more penetrating than the standard aerosols through mechanical filters (though the actual penetration was low), but are less penetrating than standard aerosols through electrostatic filters.

Any explanation for this observation is at present conjectural, but if shear flow in filters does align the fibres, it will orientate them so as to reduce interception effects, whereas the effect of electric fields is to align fibres in the field direction, and this, in filters, is usually perpendicular to the flow.

Combined effects of two or more capture mechanisms

In investigational work to validate or check a theoretical prediction, it is best to arrange filtration parameters so that one mechanism acts alone; but in a real situation several mechanisms usually act together. An exact description of filtration in this situation is not easy, because the approximations necessary to obtain an analytical expression for the description of one process and those necessary if another is to be described are often mutually exclusive.

The simplest general approach is to assume that the two processes act independently, and this can, in principle, describe any two processes. If this assumption is made, and the processes have calculated single fibre efficiencies E_1 and E_2, then a fraction $(1 - E_1)$ would escape capture by the first process if it acted alone. If these are then subjected to the second process, a fraction $(1 - E_1)(1 - E_2)$ will escape capture by both processes. This means that if E_{12} is the total single fibre efficiency,

$$E_{12} = E_1 + E_2 - E_1 E_2 \qquad (4.59)$$

and for several independent processes,

$$1 - E_{1 \ldots j} = \prod_{i=1}^{j} (1 - E_i). \qquad (4.60)$$

It is hard to visualise several processes acting together, but the expressions are symmetric with respect to all processes. If the rigorous condition is relaxed, it is often useful to write the total single fibre efficiency for two processes acting together as,

$$E_{12} = E_1 + E_2 + E'_{12}(N_1, N_2) \tag{4.61}$$

where N_1 and N_2 are the dimensionless parameters appropriate to the two processes. Equation 4.61 is completely general and is really just a definition of $E'_{12}(N_1, N_2)$, which is often negative; and it is useful only if the following inequality is satisfied,

$$|E'_{12}(N_1, N_2)| \ll E_1, E_2. \tag{4.62}$$

It is possible in some cases to describe in greater detail what happens when just two processes act together. The combined effect of interception and inertial impaction has been discussed above, and the pairs to be considered in detail below are: diffusion and interception; gravity and interception; and inertial impaction and gravity.

Diffusion and interception

The simple picture of particle capture by diffusional deposition alone can be extended to the situation where interception is included, because with the same assumptions as before it is possible to define a limiting trajectory, which ends not at the rear stagnation point but at a distance of one particle radius above the fibre surface. The equation solved previously can be solved now, and the result is,

$$E_{DR} = \frac{1}{\zeta R^2} \left[\frac{3D\zeta R^2 \pi}{2U} + \left(\frac{d_p}{2} \right)^3 \right]^{2/3}. \tag{4.63}$$

As d_p approaches zero, this reverts to the form appropriate for diffusional deposition only. Moreover when D approaches zero it becomes the value previously calculated for interception, equation 4.9, which is fortuitous since the particle would never actually reach the rear stagnation point by the process of interception alone. One particular value of equation 4.63 is that it indicates the existence of a most penetrating particle size, which is that value of d_p that gives the minimum of the expression.

A formula appropriate to a fan model filter (56, 58, 68), acting by interception and diffusion and including the effect of aerodynamic slip is,

$$E_{DR} = E_D + E_R + E'_{DR}(Pe, N_R) \tag{4.64}$$

in which E_D differs slightly from the expression in equation 4.51,

$$E_D = 2.7 Pe^{-2/3}[1 + 0.39\zeta'^{-1/3} Pe^{-1/3} Kn] \tag{4.65}$$

where, for the particular model considered, the hydrodynamic factor, including the effect of aerodynamic slip, is,

$$\zeta' = -0.5 \ln(c) - 0.52 + 0.64c + 1.43(1-c)Kn. \qquad (4.66)$$

E_R also differs from previously calculated values,

$$E_R = \frac{1}{2\zeta'} [(1+N_R)^{-1} - (1+N_R) + 2(1+N_R)\ln(1+N_R)$$
$$+ 2.86Kn(2+N_R)N_R(1+N_R)^{-1}] \qquad (4.67)$$

and the combination term is,

$$E'_{DR}(Pe, N_R) = 1.24\zeta'^{-1/2}Pe^{-1/2}N_R^{2/3}. \qquad (4.68)$$

Most penetrating particle

The existence of a size, in the sub-micrometre range (69), where the aerosol particles are most penetrating, as illustrated in Fig. 4.17, has been demonstrated by a large number of authors. In particular, the diameter has been identified as 0.68 μm at a filtration velocity of 0.001 ms^{-1} through a filter with fibre diameters of 1.5 μm, the size falling steadily to 0.5 μm as the filtration velocity was increased to 0.01 ms^{-1} (70); and as 0.15–0.20 μm for particles in an aerosol passing through a filter of mean fibre diameter 0.8 μm at a velocity of 0.025 ms^{-1} (71).

In a series of tests (72) on paper filters, the size of the most penetrating particle was observed to shift from > 1.0 μm at a filtration velocity of 0.001 ms^{-1}, to 0.6 μm at 0.01 ms^{-1}, to 0.4 μm at 0.25 ms^{-1}, in accordance with the theoretical prediction that the most penetrating size is reduced by increasing filtration velocity.

In other tests, carried out at a filtration velocity of 0.70 ms^{-1}, through a filter of the HEPA class but of unspecified fibre size, a diameter of 0.04 μm was observed for the most penetrating particles (73).

The most penetrating size calculated from equation 4.63 shows the correct behaviour pattern, and the size is close to that given by experiment, though the extent of agreement is probably fortuitous, for the equation is only approximate.

The variation of the size of the most penetrating particle has been studied (74), with fibre diameter, filtration velocity and packing fraction as variables, two being held constant and the third varied. The diameter of the most penetrating particle has been found, consistent with the data above, to increase with fibre diameter and packing fraction and to decrease with filtration velocity.

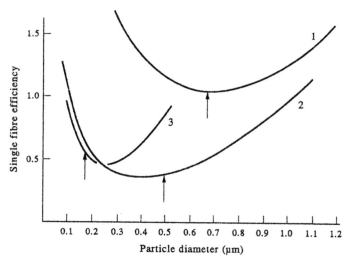

FIG. 4.17 Calculated single fibre efficiency for particle capture by interception and diffusion, according to simple theory, equation 4.63, illustrating the existence of a most penetrating particle size: (1) $U=0.001$ ms^{-1}, $d_f=1.5$ μm; (2) $U=0.01$ ms^{-1}, $d_f=1.5$ μm; (3) $U=0.025$ ms^{-1}, $d_f=0.8$ μm. Arrows indicate the experimentally observed most penetrating size (70, 71).

A study of penetration of particles in this size range, has also been made in the transition flow regime (75), where aerodynamic slip is important. The most penetrating size is observed to increase with increasing Knudsen number, though the penetration of all particles is considerably reduced as the Knudsen number increases.

A plausible explanation for the shift is that the effect of aerodynamic slip on the flow pattern would tend to increase both the interception and the diffusional capture efficiencies, but slip effects would also increase the coefficient of diffusion of the particles themselves, which would further augment diffusional capture.

Other workers (76) show the size of the most penetrating particle to decrease as the Knudsen number increases, but the change in *Kn* is brought about in these measurements by a reduction in fibre size, which could itself be responsible for this effect.

Gravity and interception

The relative importance of interception and gravity in effecting particle capture depends on the relative values of the calculated single fibre efficiencies by each process acting alone. These expressions, given in equations 4.9 and 4.31 show that an important criterion is the fibre diameter, since single fibre efficiency by interception is inversely proportional to the square of this length, but that for gravitational capture is independent of it. The results shown above for gravity acting alone (35) were obtained with filters having particularly large fibre diameters. Fine-fibred filters are much more likely to exhibit the two effects together.

A theoretical study of the combined effects of gravity and interception, based on the special properties of solenoidal force field, has been carried out (37), resulting in the expression for single fibre efficiency for a general situation, where the velocity vector of the macroscopic flow and that of gravity make an angle θ_G, which is zero for downflow, π for upflow and $\pi/2$ for horizontal flow:

$$E_{GR} = E_R(1 + 2\Lambda \cos \theta_G + \Lambda^2)^{1/2} \tag{4.69}$$

where

$$\Lambda = \frac{N_G(1 + N_R)}{E_R}. \tag{4.70}$$

Gravity combined with interception has its greatest effect when $\theta_G = 0$, the downflow condition.

$$E_{GR} = E_R + N_G(1 + N_R) \tag{4.71}$$

Where $N_G \gg N_R$ this reverts to an expression identical to that in equation 4.31 (37). When $N_R \gg N_G$ the action of gravity is only to modify slightly the pure interception result, but if upflow conditions are considered, the expression,

$$E_{GR} = E_R - N_G(1 + N_R) \tag{4.72}$$

results and so the two results can be combined to eliminate E_R, though not N_R. The difference between them should give results proportional to N_G, which had been shown previously (77).

In the situation of crossflow, where $\theta_G = \pi/2$, the expression becomes,

$$E_{GR} = [E_R^2 + N_G^2(1 + N_R)^2]^{1/2} \tag{4.73}$$

and it is clear that when $N_G \gg N_R$ this reverts to the linear dependence of E_G on N_G shown in equation 4.31. On the other hand, when the effect of gravity is a small perturbation on interception, the dependence of single fibre efficiency on N_G is quadratic.

Returning to the case when $N_G \gg N_R$, the simple proportionality is observed for every angle of flow that does not have an upflow component. When there is some measure of upflow, gravity can act to hinder deposition.

Interception, diffusion and gravity

The theory above has been extended to the situation where the three effects act together (78). The conclusion of the work is that, during upflow, when gravity is acting against the other effects, the minimum in single fibre efficiency observed as a result of the combination of interception and diffusion is shifted to a larger particle size, becoming weaker and broader.

Inertial impaction and gravity

A comparison between the capture of particles by inertial impaction and by gravity shows that whereas both vary according to the aerodynamic diameter of the particle, the velocity dependence is opposite in the two cases, which indicates that there will be some velocity at which the combined capture efficiency will be a minimum. An approximation to this velocity can be obtained from equations 4.25, 4.31 and 4.59. The velocity predicted is higher than that observed (35, 36), though agreement is reasonable in view of the fact that equation 4.25 is based on data from fibrous filters, whereas the experiments were carried out on porous foams with the structure illustrated in Figs 2.9 and 2.10.

Size-selection by interception, gravity and inertial impaction

Any of the three processes could be used for size-selection of particles. In the case of interception, particles are selected with a single fibre efficiency proportional to the square of the geometric diameter (equation 4.2), which means that the size-selection curve for an aerosol penetrating through the filter will have a Gaussian form:

$$P \sim \exp(-\alpha \, d_{\mathrm{p}}^2) \tag{4.74}$$

where α is a constant. In the case where gravity acts alone, a similar shape will be observed, though the size selection in this instance will be according not to geometric diameter but to aerodynamic diameter. In the case of inertial impaction the selection is also according to aerodynamic diameter, but the selection curve will be much sharper.

Aerodynamic size selection is particularly important because it is aerodynamic diameter that decides whether or not an inhaled particle will penetrate into the respiratory part of the lungs. This selection method can therefore be used to separate respirable from non-respirable dust as part of a sampling exercise, provided that the size-selection curve has an appropriate shape. Selection curves observed with the same filter acting in each of the two regimes are shown in Fig. 4.18 where the much sharper selection of the impaction curve is apparent. The experimentally observed selection curve for gravitational setting is less sharp than the theoretical curve.

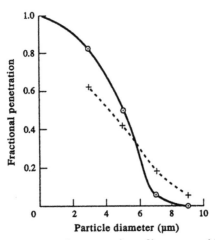

FIG. 4.18   Size selection curves for porous foam filter operated in two regimes, $\odot$, inertial impaction; $+$, gravitational settling (35). ($\copyright$ Crown Copyright; by permission of Her Majesty's Stationery Office.)

The existence of a minimum in the curve of single fibre efficiency against velocity means that there is a range of velocities over which the size selection will vary only weakly. This behaviour has been exploited in the development of a filter size selector for use during the variable airflow encountered in breathing (36).

Specification of mechanism-dependent regimes

An expression for single fibre efficiency combining all of the processes mentioned, and valid in all situations, would undoubtedly be most useful, but nothing resembling it has appeared so far and it is probably a vain pursuit. When not all individual processes admit to an analytical treatment it is optimistic to expect the sum of processes to do so.

A number of formulae fitted to theoretical calculations or experimental observations exist, but these should be used carefully, with attention to any limitation in the type of filter described, and to the range of parameters over which the expressions are valid. They may be misleading outside this range.

The best approach is to identify the dominant mechanism; and evaluation of the appropriate dimensionless parameters should help in this. When this has been done it should be possible to pick a reasonable theoretical description.

Two-dimensional maps with orthogonal directions specifying particle size and filtration velocity have been put forward (38), with specific areas identified with respect to the mechanism or mechanisms that are dominant there. Again, in a complete picture an N-dimensional map would be required, in which N is the number of fundamental parameters needed to describe the process. A two-dimensional map is a section of such a hyperspace, and is valid only for the specific values of $N-2$ parameters held constant.

Variation of deposition site with capture process

The area of fibre over which deposition takes place is an important feature of any process, becoming more important when the question of clogging is considered. It is most easily expressed in terms of the probability of particle deposition on the fibre as a function of θ, the angle between the deposition site and the forward stagnation point. This probability alters as the fibre becomes loaded, but in this chapter we are considering only clean filters.

When interception acts, deposition takes place over the half of the fibre surface facing the approaching air; and the theoretical deposition pattern, which varies as $\cos \theta$, is observed experimentally (79). When diffusional deposition occurs, capture takes place over the entire surface, though the deposition probability falls monotonically from a maximum at the forward stagnation point to a minimum at the rear (80), at a rate of decrease that is greater at small Peclet numbers.

In the case of inertial impaction, the area has an upper bound of the forward half of the fibre surface, and may be considerably less. Numerical calculations (81) have shown that the area varies not only with Stokes number, but also with interception parameter and Reynolds number. The angle between the forward stagnation point and the limiting capture position, where the particle's velocity on contact is tangential to the surface, is clearly closely related to the single fibre efficiency; and it is plotted against this parameter in Fig. 4.19. For small or moderate values of single fibre efficiency, deposition takes place over only a limited area of fibre.

The angular dependence of deposition by gravitational capture has also been studied in detail (37). Broadly, the results are that deposition

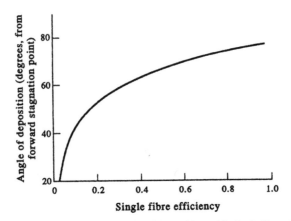

FIG. 4.19 Calculated variation of extent of deposition with single fibre efficiency for capture by inertial impaction. Extent of deposition is measured as the angle from the forward stagnation point, over which deposition occurs (81). (© 1987, Elsevier Science Publishing Co. Inc.)

takes place over roughly half of the fibre surface, the exact location of which depends on the angle, θ_G, between the flow direction and the direction in which gravity acts.

Effect of leakage on filter performance

That a leak or a pinhole will increase aerosol penetration through a filter is obvious; and a number of generalisations will be given, which should give some insight into the performance of leaky filters.

A leak will exhibit a measure of size-selection in broadly the same way that a filter does. Since the air passing through a leak does so faster than the air passing through the filter, there will be a convergence of streamlines into the leak, and this will present an opportunity for inertial losses, rather analogous to the under-sampling of non-isokinetic samplers (82). Within the leak there is an opportunity for impaction, settling and diffusional deposition. The best clues to the mechanism acting are the dependences of leakage on particle size and on filtration velocity, both of which indicate that inertial impaction is the primary means by which leaks fail to transmit particles. Typical leakage curves are shown in Fig. 4.20 (83).

Particles smaller than 1.0 μm tend to penetrate highly, though some diffusional deposition does occur. However, fine particle leakage also has a velocity-dependent component, due to the fact that a smaller fraction of air passes through the leak or pinhole at higher velocities (84). This effect is illustrated in Fig. 4.21 (85), which shows a reduction in penetration of aerosol through a leaky filter at high velocity, in

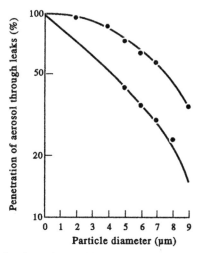

F<small>IG.</small> 4.20 Particle size dependence of leakage with a filter of moderate efficiency. The two graphs illustrate extremes observed over a number of measurements (83). (By permission of the American Industrial Hygiene Association.)

contrast with the increase in penetration through an intact filter caused by a reduction of diffusional deposition.

Potential problems with model filters

The important link that model filters give between theory and experiment, requires that they obey the normal filtration equations given in Chapter 1. For this to be the case, it is important that the pattern of particle depletion caused by one layer of fibres should not correlate with the structure of the layers downstream of it, as it might in model filters of the regular structures shown in Fig. 3.5. If, however, the diffusivity of the particles is sufficiently great, the aerosol will become uniform in concentration within an inter-layer spacing, $2e$, and so the problem will not arise. The criterion for this is,

$$\frac{4\,De}{U\,d_f^2 E_s^2} \gtrsim 1. \qquad (4.75)$$

On the other hand the problem may be overcome if the fibres are arranged effectively at random (86) so that the approaches of the aerosol to every fibre may be regarded as independent events. A mathematical theory (87) seeks to avoid such problems by circumventing the single fibre approach.

If particles have a finite probability of bouncing from fibres that they strike, it is necessary that a particle after bouncing recovers the velocity

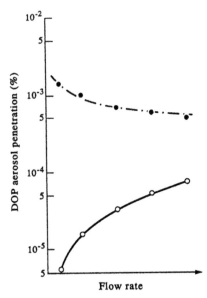

FIG. 4.21 Velocity dependence of penetration of a submicrometre aerosol through a high efficiency filter: ○, through an intact filter; ●, through a leaky filter (85). (By permission of M. Furuhashi.)

of the carrying air within an inter-layer distance. A detailed description of bouncing is given in Chapter 7, but all that is required here is that the stopping distance of the particle, given by equation 4.16, should be significantly smaller than the inter-layer spacing, which is summarised by (88):

$$c \lesssim \frac{1}{St^2}.$$ (4.76)

References

1. KUWABARA, S. The forces experienced by randomly distributed parallel cylinders or spheres in viscous flow at small Reynolds numbers. *J. Phys. Soc., Japan*, 1959, **14**, 527–532.
2. DAVIES, C. N. *Air Filtration*, Academic Press, London, 1973.
3. LEE, K. W. and GIESEKE, J. A. Note on the approximation of interceptional collection efficiencies. *J. Aerosol Sci.*, 1980, **11**, 335–341.
4. FARDI, B. and LIU, B. Y. H. Efficiency of fibrous filters with rectangular fibres. *Aerosol Sci. Technol.*, 1992, **17**, 45–58.
5. DORMAN, R. G. *Dust Control and Air Cleaning*, Pergamon Press, Oxford, 1974.
6. BROWN, R. C. and WAKE, D. Air filtration by interception — theory and experiment. *J. Aerosol Sci.*, 1991, **22**(2), 181–186.
7. PICH, J. The effectiveness of the barrier effect in fibre filters at small Knudsen numbers. *Staub Reinhalt. Luft.*, 1966, **26**, 1–4.
8. WAKE, D. and BROWN, R. C. Measurements of the filtration efficiency of nuisance dust masks against respirable and non-respirable aerosols. *Ann. Occ. Hyg.*, 1988, **32**(3), 295–315.
9. FUCHS, N. A. *The Mechanics of Aerosols*, p. 70ff, Pergamon, Oxford, 1964.

10. CLIFT, R., GRACE, J. R. and WEBER, M. E. *Bubbles, Drops and Particles*, pp. 286–291, Academic Press, New York, 1978.

11. AYERS, F. *Theory and Problems of Differential Equations*, p. 188ff, Schaum, New York.

12. GOREN, S. L. and O'NEILL, M. E. On the hydrodynamic resistance to a particle of a dilute suspension when in the neighbourhood of a large particle. *Chem. Eng. Sci.*, 1971, **26**, 325–338.

13. GOREN, S. L. The hydrodynamic force resisting the approach of a sphere to a plane in slip flow. *J. Coll. Int.*, 1973, **44**(2), 356–360.

14. INGHAM, D. B., HILDYARD, L. T. and HILDYARD, M. L. On the critical Stokes number in potential and viscous flow near bluff bodies. *J. Aerosol Sci.*, 1990, **21**(7), 935–946.

15. STECHKINA, I. B., KIRSCH, A. A. and FUCHS, N. A. Studies on fibrous aerosol filters — IV. Calculation of aerosol deposition in model filters in the region of maximum penetration. *Ann. Occ. Hyg.*, 1969, **12**, 1–8.

16. BROWN, R. C. Porous foam size selectors for respirable dust samplers. *J. Aerosol Sci.*, 1980, **11**, 151–159.

17. HARROP, J. A. and STENHOUSE, J. I. T. Theoretical predictions of inertial impaction efficiency in filters. *Chem. Eng. Sci.*, 1969, **24**, 1475–1489.

18. NGUYEN, X. and BEECKMANS, J. M. Single fibre capture efficiency of aerosol particles in real and model filters in the inertial interception domain. *J. Aerosol Sci.*, 1975, **6**, 205–212.

19. TSIANG, R. C. and TIEN, C. Trajectory calculation of particle deposition in model filters consisting of parallel fibres. *Can. J. Chem. Eng.*, 1981, **59**, 595–600.

20. STENHOUSE, J. I. T., HARROP, J. A. and FRESHWATER, D. C. The mechanisms of particle capture in gas filters. *Aerosol Sci.*, 1970, **1**, 41–52.

21. MCLAUGHLAN, C., McCOMBER, P. and GAKWAYA, A. Numerical calculation of particle collection by a row of cylinders in a viscous fluid. *Can. J. Chem. Eng.*, 1986, **64**, 205–210.

22. SUNEDA, S. K. and LEE, C. H. Aerosol filtration by fibrous filters at intermediate Reynolds numbers (< 100). *Atmos. Environ.*, 1974, **8**, 1081–1084.

23. EMI, H., OKUYAMA, K. and ADACHI, M. The effect of neighbouring fibres on the single fibre inertia-interception efficiency of aerosols. *J. Chem. Eng., Japan*, 1977, **10**(2), 148–153.

24. LANDAHL, H. D. and HERMANN, R. G. Sampling of liquid aerosols by wires, cylinders and slides, and the efficiency of impaction of droplets. *J. Coll. Sci.*, 1949, **4**, 103–136.

25. THOM, A. The flow past circular cylinders at low speeds. *Proc. Roy. Soc.*, 1933, **A141**, 651–669.

26. KIRSCH, A. A. and STECHKINA, I. B. Inertial deposition of aerosol particles in model filters at low Reynolds numbers. *J. Aerosol Sci.*, 1977, **8**, 301–307.

27. MIYAGI, T. Viscous flow at low Reynolds number past an infinite row of equal circular cylinders. *J. Phys. Soc., Japan.*, 1958, **13**, 493–496.

28. EMI, H., WANG, C. S. and TIEN, C. Transient behaviour of aerosol filtration in model filters. *AIChE J.*, 1982, **28**(3), 397–405.

29. BREWER, J. M. and GOREN, S. L. Evaluation of metal oxide whiskers grown on screens for use as aerosol filtration medium. *Aerosol Sci. Technol.*, 1984, **3**(4), 411–429.

30. LEE, K. W. and LIU, B. Y. H. Experimental study of aerosol filtration in fibrous filters. *Aerosol Sci. Technol.*, 1982, **1**, 35–46.

31. FAN, K. C., WAMSLEY, B. and GENTRY, J. W. The effect of Stokes and Reynolds numbers on the collection efficiency of grid filters. *J. Coll. Int. Sci.*, 1978, **65**(1), 162–173.

32. GENTRY, J. W. and CHOUDHARY, K. R. Collection efficiency and pressure drop in grid filters of high packing density at intermediate Reynolds numbers. *J. Aerosol Sci.*, 1975, **6**, 277–290.

33. STENHOUSE, J. I. T., BROOM, G. P. and CHARD, N. T. J. High inertia fibrous filtration — optimisation conditions. *Proc. Filtration Soc.*, 1978, 128–136.

34. PTAK, T. and JAROSZCYK, T. Theoretical—experimental aerosol filtration for fibrous filters at intermediate Reynolds numbers. *Proceedings of the 5th World Filtration Congress*, 1989, 566–572.

35. WAKE, D. and BROWN, R. C. Filtration of monodisperse aerosols and polydisperse dusts by porous foam filters. *J. Aerosol Sci.*, 1991, **22**(6), 693–706.

36. GIBSON, H. and VINCENT, J. H. Penetration of dust through porous foam filter media. *Ann. Occ. Hyg.*, 1981, **24**(2), 205–215.

37. PICH, J. Theory of gravitational capture of particles in fibrous aerosol filters. *Aerosol Sci.*, 1973, **4**, 217–226.

38. EMI, H., OKUYAMA, K. and YOSHIOKA, N. Prediction of collection efficiency of aerosols by high porosity fibrous filter. *J. Chem. Eng., Japan*, 1973, **6**(4), 349–354.

39. DUCHIN, S. S. and DERYAGUIN, B. V. Towards a method of calculating the precipitation of disperse particles from a flow interrupted by obstacles. *Kolloid Zh.*, 1958, **20**, 326–328 (in Russian). HSE Translation 7221 (1977).

40. ALLEN, H. S. and MAXWELL, R. S. *A Textbook of Heat*, Macmillan, London, 1962.

41. CUNNINGHAM, E. On the velocity of steady fall of spherical particles through a fluid medium. *Proc. Roy. Soc.*, 1910, **83A**, 357–365.

42. MILLIKAN, R. A. Coefficient of slip in gases and the law of reflections of molecules from the surfaces of solids and liquids. *Phys. Rev.*, 1923, **21**(3), 217–238.

43. ALLEN, M. A. and RAABE, O. G. Slip correction measurement of spherical solid aerosol particles in improved Millikan Apparatus. *Aerosol Sci. Technol.*, 1985, **4**(3), 269–286.

44. CHENG, Y. S., ALLEN, M. D., GALLEGOS, D. P., YEH, H. C. and PETERSON, K. Drag force and slip correction of aggregate aerosols. *Aerosol Sci. Technol.*, 1988, **8**, 199–214.

45. KAY, J. M. *An Introduction to Fluid Mechanics and Heat Transfer*, C. U. P., Cambridge, 1963.

46. GORMLEY, P. G. and KENNEDY, M. Diffusion from a stream flowing through a cylindrical tube. *Proc. Roy. Ir. Acad.*, 1949, **52**, 163–169.

47. NATANSON, G. L. Diffusional precipitation of aerosols on a streamlined cylinder with a small capture coefficient. *Proc. Acad. Sci., USSR. Phys. Chem. Section*, 1957, **112**, 21–25.

48. PICH, J. A note on the diffusive deposition of aerosols on a cylinder. *Aerosol Sci.*, 1970, **1**, 17–19.

49. LEE, K. W and LIU, B. Y. H. Theoretical study of aerosol filtration in fibrous filters. *Aerosol Sci. Technol.*, 1982, **1**, 147–161.

50. STECHKINA, I. B. and FUCHS, N. A. Studies on fibrous aerosol filters — I. Calculation of diffusional deposition of aerosols in fibrous filters. *Ann. Occ. Hyg.*, 1966, **9**, 59–64.

51. RAO, N. and FAGHRI, M. Computer modelling of aerosol filtration by fibrous filters. *Aerosol Sci. Technol.*, 1988, **8**(2), 133–156.

52. MASLIYAH, J. H. Aerosol removal by diffusion and interception on mats of elliptic fibres. *Can. J. Chem. Eng.*, 1975, **53**, 568–571.

53. BANKS, D. O. and KUROWSKI, G. I. Diffusion deposition on a cylinder due to nearly parallel flow. *Aerosol Sci. Technol.*, 1988, **8**(2), 189–196.

54. CHENG, Y. S., YAMADA, Y. and YEY, H. C. Diffusion deposition in model fibrous filters with intermediate porosity. *Aerosol Sci. Technol.*, 1990, **12**(2), 286–299.

55. CHENG, Y. S., YEH, H. C. and YAMADA, Y. Diffusion deposition on model fibrous filters with intermediate porosity. *Proceedings of 3rd International Aerosol Conference, Kyoto*, Sept. 24–27, 1990, 691–694, Pergamon.

56. KIRSCH, A. A. and CHECHUEV, P. V. Diffusion deposition of aerosol in fibrous filters at intermediate Peclet numbers. *Aerosol Sci. Technol.*, 1985, **4**, 11–16.

57. PICH, J. The filtration theory of highly dispersed aerosols. *Staub Reinhalt. Luft.*, 1965, **5**, 16–23 (in English).

58. KIRSCH, A. A., STECHKINA, I. B. and CHECHUEV, P. V. Precipitation of submicron aerosol particles in fibrous filters at high temperatures. *Theor. Found. Chem. Eng.*, 1988, **22**(4), 379–384.

59. STERN, S. T., ZELLER, H. W. and SHECKMAN, A. I. The aerosol efficiency and pressure drop of a fibrous filter at reduced pressures. *J. Coll. Sci.*, 1960, **15**(6), 546–562.

60. GANS, R. Zur theorie der Brownschen molekularbewegung. *Annalen der Physik*, 1928, **86**, 628–656 (in German).

61. JEFFERY, G. B. The motion of ellipsoidal particles immersed in a viscous fluid. *Proc. Roy. Soc.*, 1923, **102A**, 161–179.

62. TREVELYAN, B. J. and MASON, S. G. Particle motion in sheared suspensions. *J. Coll. Int. Sci.*, 1951, **6**, 354–367.

63. BENARIE, M. The influence of the form of solid dust particles on their filterability. *Staub Reinhalt. Luft.*, 1963, **23**(2), 50–51 (in English).

64. FU, T. H., CHENG, M. T. and SHAW, D. T. Filtration of chain aggregate aerosols by model screen filters. *Aerosol Sci. Technol.*, 1990, **13**, 151–161.

65. CHERRIE, J. W. An interim report on the measurement of the efficiency of respirator filters against very fine man-made mineral fibres. *I.O.M. Report BP.31075/2/D.*

66. ORTIZ, L. W., SOVERHOLM, S. C. and VALDEZ, F. O. Penetration of respirator filters by an asbestos aerosol. *Am. Ind. Hyg. Assoc. J.*, 1988, **49**(9), 451–460.

67. BROSSEAU, L. M., ELLENBECKER, M. J. and EVANS, J. S. Collection of silica and asbestos aerosols by respirators at steady and cyclic flow. *Am. Ind. Hyg. Assoc. J.*, 1990, **51**(8), 420–426.

68. KIRSCH, A. A. and STECHKINA, I. B. The theory of aerosol filtration with fibrous filters, In *Fundamentals of Aerosol Science* (ed. D. T. SHAW) p. 165, Wiley, New York, 1978.

69. STAFFORD, R. G. and ETTINGER, H. J. Filter efficiency as a function of particle size and velocity. *Atmos. Env.*, 1972, **6**, 353–362.

70. THOMAS, J. W. and YODER, R. E. Aerosol size for maximum penetration through fiberglass and sand filters *AMA Arch. Ind. Health*, 1956, **13**, 545–549.

71. DYMENT, J. Use of a Goertz aerosol spectrometer for measuring the penetration of aerosols through filters as a function of particle size. *Aerosol Sci.*, 1970, **1**, 53–67.

72. SINCLAIR, D. Penetration of HEPA filters by submicron aerosols. *J. Aerosol Sci.*, 1976, **7**, 175–179.

73. RIMBERG, D. Penetration of IPC1478, Whatman 41 and type 54 filter paper as a function of particle size and velocity. *Am. Ind. Hyg. Assoc. J.*, 1969, **30**, 394–401.

74. LIU, B. Y. H. and RUBOW, K. L. *Air Filtration by Fibrous Media. Fluid Filtration: Gas* (ed. R. R. RABER), Vol. l, pp. 1–12, *American Society for Testing and Materials, Philadelphia*, 1986.

75. ZHANG, Z. and LIU, B. Y. H. Experimental study of aerosol filtration in the transition flow regime. *Aerosol Sci. Technol.*, 1992, **16**, 227–235.

76. PATEL, S. N., YAMAMOTO, H., MASUDA, S. and SHAW, D. T. Filters for diesel particulate control. *Advances in Filtration and Separation Technology* (ed. K. L. RUBOW), Vol. 4, pp. 252–257, *American Filtration Society*, 1991.

77. THOMAS, J. W., RIMBERG, D. and MILLER, T. J. Gravity effect in aerosol filtration. *Aerosol Sci.*, 1971, **2**, 31–38.

78. PICH, J. and SPURNY, K. Direction of fluid flow and the properties of fibrous filters. *Aerosol Sci. Technol.*, 1991, **15**, 179–183.

79. GILLESPIE, T. On the adhesion of drops and particles on impact at solid surfaces. *J. Coll. Sci.*, 1955, **10**, 266–280.

80. KANAOKA, C., EMI, H. and TANTHAPANICHAKOON, W. Convective diffusional deposition and collection efficiency of aerosol on a dust loaded fibre. *AIChE J.*, 1983, **29**(6), 895–901.

81. WANG, P. K. and JAROSZCZYSCK, Y. The grazing collision angle of aerosol particles colliding with infinitely long cylinders. *Aerosol Sci. Technol.*, 1991, **15**, 149–155.

82. VINCENT, J. H. *Aerosol Samplers: Science and Practice*, Wiley, New York, 1989.

83. HINDS, W. C. and KRASKE, G. Performance of dust respirators with facial seal leaks 1. Experimental. *Am. Ind. Hyg. Assoc. J.*, 1987, **48**(10), 836–841.

84. FAHRBACH, I. J. The effect of leaks on total penetration velocity and concentration at perforated filters. *Staub Reinhalt. Luft.*, 1970, **30**(12), 45–52.

85. FURUHASHI, M. and KOZUKA, M. Pinhole effect in air filter and aerosol penetration rate. *Proceedings 7th International Symposium on Contamination Control, Paris, Sept. 18–21,* 1984.

86. OVERCAMP, T. J. Filtration by randomly distributed fibres. *J. Aerosol Sci.*, 1985, **16**(5), 473–475.

87. SHAPIRO, M. and BRENNER, H. Dispersion/reaction model of aerosol filtration by porous filters. *J. Aerosol Sci.*, 1990, **21**(1), 97–125.

88. DAVIES, C. N. The retention of particles in filters. *J. Aerosol Sci.*, 1985, **16**(5), 473–475.

CHAPTER 5

Electrically Charged Filter Material

Introduction

Electrically charged filter material has a history of several decades; in fact the first such material was used for a period of years before its mechanism of action was properly understood. The advantage of materials of this type is that the charge on the fibres considerably augments the filtration efficiency without making any contribution to the airflow resistance. Several materials carrying permanent electric charge now exist, finding wide use in situations where a high efficiency is required along with a low resistance, such as in respirator filters.

Earlier chapters have shown that the single-fibre capture efficiency, by interception, of micrometre-sized particles by a 20 μm diameter fibre (a typical size for animal or vegetable fibres, or for synthetic fibres that can be processed with normal textile equipment) is extremely low; and an example of this low efficiency is shown as curve 1 in Fig. 5.1. The efficiency of the same material, with a level of charge on the fibres typical of electrically charged material, is considerable, as curve 2 illustrates.

Basic mechanisms of action

Electrically charged fibres attract both charged and neutral particles. The detailed theory will be given in Chapter 6, but the mechanism of action is illustrated in Fig. 5.2. The capture of oppositely charged particles by coulomb forces is self-evident. The capture of neutral particles comes about by the action of polarisation forces. The electric field of the fibre induces a dipole in a neutral particle, or indeed a charged one, and then attracts it. The strength of the induced dipole depends upon the volume of the particle and the dielectric constant of its constituent material. The component that is nearer to the fibre

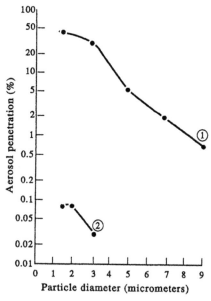

FIG. 5.1 Performance of electrically charged and electrically neutral filters of identical geometry: (1) neutral material (the capture efficiency for 5 μm particles is relatively high. That for 1 μm particles is very low); (2) charged material.

experiences an attractive force slightly larger than the repulsive force acting on its counterpart, since the electric field diminishes with distance from the fibre; and this slight imbalance causes the particle to be attracted, whatever the sign of charge on the fibre. The efficiency of this mechanism, and that of the coulomb interaction, depends on the quotient of the drift velocity of the particle under the influence of the electric force tending to attract it to the fibre and the convective velocity of the flow field tending to take it past. All electrostatic filters are, therefore, more efficient at low filtration velocities.

Classification of material

Electrically charged filter materials may be classified in a number of ways, the most straightforward being based on the manner in which material becomes electrically charged. There are three principal charging processes: triboelectric charging, corona charging and charging by induction. A further process, the freezing-in of charge, has been demonstrated experimentally, but does not appear to be used in commercial manufacture.

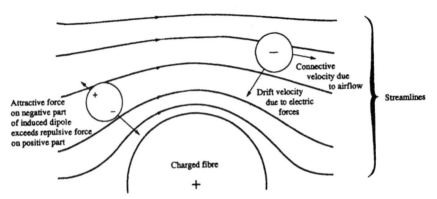

FIG. 5.2   Illustration of the capture of charged and neutral particles by a charge filter fibre. (© Crown Copyright; by permission of Her Majesty's Stationery Office.)

Triboelectrically charged material

The exchange of electric charge between two contacting materials is well known, if incompletely understood. The charging of a plastic comb on vigorous contact with human hair is a well-worn party trick, as is that of charging a balloon by contact with knitted woollen fabrics; and the development of electric charge on a rubbed amber rod, first observed by the ancients, has illuminated past generations of school-children. On a microscopic level triboelectrification can be used to generate electric charge within a filter.

Resin wool material

The first recognised electrostatic filter, the resin-wool or Hansen filter (1) used a charging process similar to the last mentioned above. Amber is fossilised resin, and particles of resin in its native form, when carded with fibres of wool, exchange charge so that the resin becomes negatively charged and the wool positively charged (2, 3).

Resin is an extremely good insulator, and its low conductivity is sufficient to ensure that the charge on the filter material is stable. Wool can be considered to be a conductor. Under conditions of ambient relative humidity it has a low electrical resistance, and this antistatic property is one of the reasons why wool is such a comfortable fabric to wear. As a conductor, it will develop whatever charge is necessary to reduce the electrostatic energy of the system to a minimum; and the configuration that achieves this takes the form of charges that are the electrical images of the electrostatic charges of the resin, as shown in Fig. 5.3a. This and the other components of Fig. 5.3 illustrate the charge configuration that would be predicted if the charging mech-

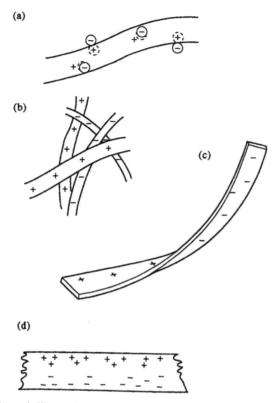

FIG. 5.3   Schematic illustration of the charge configuration on electrically charged filter material: (a) resin wool material (single fibre); (b) mixed fibre material (several fibres); (c) split fibre electret material (single fibre); (d) electret material charged in felt form (entire material).

anism described took place without the complication of other charge-altering processes. It will be shown later in the chapter, and in Chapter 6, that this is an over-simplification; and so the figures should be regarded as average charge configurations.

An electron micrograph of resin wool material is shown in Fig. 5.4. It is important to realise that the wool will not conduct away the charge on the resin, because any charge on part of the resin not in contact with the wool would have to pass through an insulator (the resin) to reach the conductor. This point is essential for the proper understanding of charge stability in electrostatic filters.

What the presence of a conductor does mean is that on a macro-scopic level the charge on the material is low. It contains positive and negative electric charge almost in balance, a property that can be verified by putting a sample of the material into a Faraday ice pail connected to an electrometer. The result of this is a very low positive

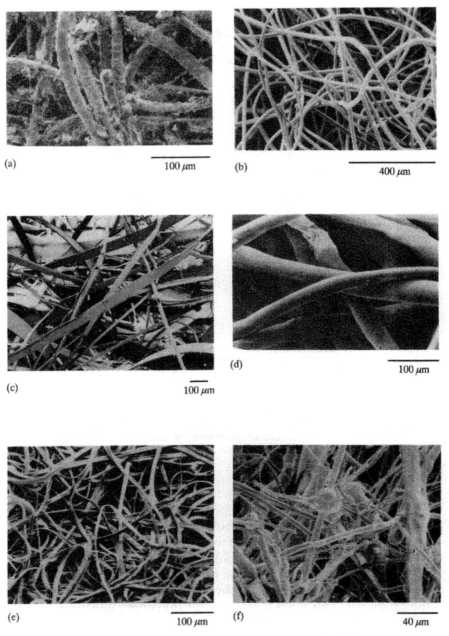

(a)   100 μm   (b)   400 μm

(d)   100 μm

(c)   100 μm

(e)   100 μm   (f)   40 μm

FIG. 5.4 Electron micrographs of electrically charged filter material. (I am indebted to Mr G. S. Revell of the Health and Safety Executive, Research and Laboratory Services Division, for taking these and most of the other electron micrographs): (a) resin wool material; (b) mixed fibre material; (c) split fibre electret material; (d) electret material charged in felt form; (e) electrostatically spun material spun from a solution; (f) electrostatically spun material spun from a melt.

reading on the electrometer (4), the reason for which will become clear later in this section.

Mixed-fibre material

A second type of material that develops its charge by triboelectric means is material made from a binary mixture of polymer fibres (5, 6). If two yarns of dissimilar polymers are rubbed together, they will exchange charge in a consistent way so that one species will develop a positive charge and the other a negative. In most cases the charge is low and fugitive. Sometimes the charge is so unstable that it decays before a measurement can be made; and in these circumstances conservation of charge must be used to infer the sign of charge developed, from observation of the charge on the companion fibre. Measurements of this sort can be used to produce a triboelectric series like that shown in Table 5.1, defined so that fibres of a type high in the series develop a positive charge when rubbed against those below them (7).

TABLE 5.1
Triboelectric series of textile yarns

Positive

Wool
Hercosett wool[a]
Nylon 66
Nylon 6
Silk
Regenerated cellulose
Cotton
Polyvinyl alcohol
Chlorinated wool
Cellulose triacetate
Calcium alginate
Acrylic
Cellulose diacetate
Polytetrafluoroethylene
Polyethylene
Polypropylene
Poly(ethylene terephthalate)
Poly(butylene terephthalate)
Modacrylic
Chlorofibre

Negative

[a] Hercosett wool is shrink-proof wool, which is first chlorinated, and then coated with nylon as a second process.

Different observers do not quite agree on the series, probably because generic names are used to describe classes of polymers, the members of which are not chemically identical; but in general wool is the most electropositive fibre (which is the reason why any spurious charge shown by the ice-pail experiment described above is likely to be positive). Nylons are slightly less electropositive, followed by cellulose-based polymers and polyolefins in that order, with aromatic and halogenated fibres at the negative end of the series.

Though the triboelectric series gives an indication of the sign of the charge developed by each fibre type of a pair, it gives no indication of the level of charge, nor of its stability, two critical features of filter performance. The ideal mixed fibre filter, and the one now in use, consists of polypropylene and chlorinated polymer fibres. The mixed-fibre material, like the resin wool, has high charge on a microscopic level and much lower charge on a macroscopic level, and ice-pail experiments show that it usually has a small spurious negative charge, since its component fibres are low in the triboelectric series. An illustration of the charge distribution on the fibres is shown in Fig. 5.3b, and an electron micrograph is shown in Fig. 5.4b.

The method of triboelectric charging can be applied to materials after they have been given the structure of filters (8), the presence of charge after contact with a dissimilar material being apparent both in direct measurements of electric charge and in measurements of filtration efficiency. However, the improvement in efficiency brought about by this technique, though easily measurable, tends to be rather small.

Corona charged material

A common means of applying an electric charge to a polymer is that of a corona discharge. A point electrode at a high potential emits ions of its own sign, and these will drift under the influence of the electric field caused by the electrode to a collecting surface of lower potential. If the surface is that of an insulator the ions will be collected but not lost to earth, and so a static charge will develop. If the insulator is a thin sheet of polymer resting on a conductor, a compensating charge of the opposite sign will develop as the polymer is stripped away from its backing. In whatever way the polymer is charged it will rapidly develop charge in a sheet dipole configuration; and in such a form it is often termed an electret. Strictly speaking this term is incorrect; true electrets are polymers with a frozen-in electric polarisation, their form being analogous with that of a magnet, just as the word "electret" is analogous with the word "magnet". However the term "electret" is now so widely applied to material with real rather than polarisation charge that it can be considered to be sanctioned by idiom. It is, unfortunately, also now applied to polymers with any configuration of charge; and this is confusing.

Split-fibre material

Polypropylene forms a good electret, and it has a further useful property with respect to filter manufacture. If a sheet of the polymer is stretched it suffers considerable molecular realignment, becoming strong along the direction of stretching and weak in the direction perpendicular to it. This allows the sheet to be split into fibres, the process being termed fibrillation, and these fibres, which now have a line-dipole configuration of charge, can be carded and made into a filter, again with a high microscopic level of charge and a low macroscopic level.

The process by which the material is manufactured (9, 10, 11, 12) involves the polymer material being subjected to a positive corona on one face and a negative on the other. An illustration of the conjectured charge configuration is shown in Fig. 5.3c, and an electron micrograph of the material is shown in Fig. 5.4c, from which the rectangular shape of the cross-section of the fibres, along with the absence of crimp, is clear.

Material charged as a whole

The corona-charging process can be applied not only to a polymer sheet, but also to a filter material, whether this has been produced by melt-blowing (*vid inf*) or by carding and felting. An electron micrograph of an example of the latter (13) is shown in Fig. 5.4d. The corona is applied over a much smaller area when the filters rather than the polymer sheet are charged and so, although the manufacture process will be simpler, one would expect the resultant charge level to be lower as illustrated in Fig. 5.3d.

Alternatively the dipolar charge can be imparted by freezing in a polarisation charge after the material has been felted (14). The material resulting from this will be a true electret consisting of polarised fibres.

The dipole moments of the fibres produced by either of the above processes, in which whole materials are charged or polarised, will be aligned, giving the material itself a net dipole moment. Those fibres produced from a polarised and fibrillated sheet that retain a line-dipole configuration during carding will be randomly orientated, imparting no net dipole moment to the material.

Material charged by induction

Induction of charge is the third form of electric charging mechanism exploited in the production of electrically charged filter material, the process being similar to that used in the production of electrically charged sprays. A conductor placed in an electric field will develop a surface charge that will reduce its own internal field to zero. If the conductor is isolated the charge developed will be dipolar in form, but if it is attached to another body and subsequently detached, a unipolar charge may be imparted. This charged body can then move under the influence of the electric field that induced the charge in the beginning, and this describes the behaviour of the small droplets forming the spray in the process of electrostatic spraying.

Electrostatic spraying involves the production of charged liquid particles, and the process of detachment occurs simultaneously with the process of charging (15). This can be conceived as the extrusion of a filament of charged liquid which breaks up into droplets under the influence of surface tension (16). If, however, this liquid is replaced by a polymer solution or melt, the high viscosity will delay its fragmentation, and under appropriate conditions the polymer filament may solidify into a fibre before this can occur. When such a fibre eventually becomes detached, it may be deposited on the collection electrode. An accumulation of such fibres will produce a structure rather like an air-laid felt.

One effect of the process is the production of highly attenuated fibres. Conventional fibre-spinning rarely produces fibres with diameters much less than 20 μm, whereas electrostatically extruded fibres may have diameters of 2 or 3 μm or less. Figures 5.4e and 5.4f show fine fibres produced by electrostatic extrusion of a solution of polycarbonate (17, 18) and fibres produced by electrostatic extrusion of a melt of polypropylene (19). The fibres produced from the melt have a circular cross section and the material contains occasional large spheres of polymer, such inclusions being known as "shot". The solvent spun fibres have a characteristic bilobar cross section, probably caused by the outer layer solidifying first, resulting in a higher surface area/volume ratio than that of a cylindrical fibre.

The theory of droplet charging during electrostatic spraying (20, 21) may require some modification for it to be applicable to the induction charging of fibres during electrostatic extrusion, but it would certainly predict that the fibres would have unipolar charge of the same sign as that of the electrode from which they arise. Measurements on the material invariably indicate the presence of charges of both signs in almost equal amounts, indicating that there must be some other charging process in operation.

The electrostatic action of filters produced by this means is rather weaker than that of filters produced by triboelectrification or corona charging; and the fibre-laying process of manufacture allows only a limited control of structure. On the other hand, their mechanical capture efficiency is relatively high, complementing the electrostatic action.

Air currents in electric charging

Fine fibred polymers may be produced by attenuating spun fibres with the aid of an air jet (22), and this process can result in a fairly large amount of spurious charge, which may develop from contact between the melt and the spinneret or as a result of manipulation of the filter medium shortly after it is formed.

It is sometimes mentioned that assembled filters develop charge from the air passing through them during use. They may become charged indirectly as a result of this process, but it is not directly instrumental in the charging, and cannot be so because of the energies involved. If a filter fibre is to become charged by contact, energy must be available for the production of both this charge and an equal opposite charge on the body exchanging charge with it. If the latter were a solid the energy involved would be small, but if a molecule of gas in air were to impart the charge, it would become ionised as a result, requiring the ionisation energy, which is of the order of thirty electron volts, to be supplied. Under normal conditions there is no mechanism whereby this sort of energy can be imparted to a single molecule. However, fibre–fibre contact, or contact between the filter and its holder caused by the air motion, may easily result in triboelectric charging of the filter.

Non-fibrous electrostatic filter

The production, on an experimental level, of a material consisting of an electret sheet and a sheet of conductor, both aligned parallel to the airflow (23, 24, 25), illustrates that filtration can be achieved with other geometries than fibres. The electret sheet is planar and the conductor is corrugated, the two being wound together to form the structure

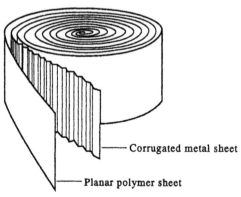

Corrugated metal sheet

Planar polymer sheet

FIG. 5.5   Non-fibrous permanently charged electrostatic material (the planar sheet is an electret, the corrugated sheet is metallic) (24). (© Crown Copyright; by permission of Her Majesty's Stationery Office.)

illustrated in Fig. 5.5, which is a system of pores in which the electric field is everywhere perpendicular to the flow direction. The filter also shows that a conductor can form an essential part of an electrostatic filter with a stable charge, a point demonstrated clearly but less vividly by the triboelectrically charged fibrous filters with their relatively high-conductivity components.

The alignment of the field means that, in principle, the entirety of an aerosol can be precipitated, whereas in a fibrous filter acting by a normal depth-filtration process, the penetration decreases exponentially but always remains finite. Experimental observations show that the non-fibrous material has a high efficiency against charged particles, combined with a low resistance to airflow. Unfortunately, the capture efficiency for neutral particles is relatively weak, since the electric field within it (which can be calculated by means of a self-consistent field calculation (24)) is, unlike that in a fibrous filter, almost uniform close to the polymer.

The importance of electric charge configuration

The effect of the quantity of charge held by a filter on its filtration efficiency is quite clear. The greater the charge, the greater will be the electric field produced, and the greater the particle capture efficiency by electric forces.

Charge configuration is equally important, though its significance is not so obvious. If all of the fibres of a filter hold charge of the same sign, the field outside the filter might be very high, but the field between two fibres carrying charge of the same sign may be low, and at points of symmetry it will be zero. The field outside the filter will not contribute to the filtration performance, but dielectric breakdown of the air will limit this field and, therefore, the magnitude of charge in this configuration that the filter could hold. As a result uniform charge is not of great value in filtration.

At the other extreme is charge that has very rapid spatial variation. A solution of the governing equation indicates that the range of electric field caused by a spatially varying charge is comparable with the period of variation. The extreme case is that of an ionic crystal like sodium chloride, which carries an enormous charge, but in such a way that equal and opposite charges are separated by atomic distances. The electric field produced will extend no further. An ionic crystal gives no indication, on a macroscopic level, of being electrically charged at all.

To be effective in air filtration electric fields must extend a significant distance beyond the surface of a charged fibre, and so the electric charge must have a spatial variation not much smaller than fibre or inter-fibre dimensions. The interpretation of measurements of electric charge must be made whilst the significance of configuration is borne in mind.

Measurement of filter charge

The most obvious way of measuring the electric charge on a filter is to place a sample in a Faraday ice-pail connected to an electrometer. If this procedure is followed for any of the materials described above, the result is that a very low level of charge is recorded. This behaviour could be anticipated from simple physical examination of the materials, because they show no obvious sign of being electrically charged. In textile vocabulary they would be described as "dead". The charge that they contain consists of positive and negative charge almost completely in balance, and the ice-pail measurement gives only the spurious excess of charge of one sign, which, as described above, is of little significance in filtration. A more significant parameter is the absolute magnitude of both the positive and the negative charge on the material.

Measurement by ionising radiation

The charge can be destroyed, in a controlled manner, by ionising radiation of known dose rate, whilst the filter performance is being measured (2,4), and as a result the magnitude of charge of each sign originally held can be calculated. The strength of an X-ray beam is normally quantified in terms of the number of ions that it produces (26), a radiation dose of 1 Roentgen resulting from a beam that produces 3.336×10^{-4} coulombs of charge of each sign, in the form of ions, in 1 m^3 of air at NTP. The same high ionisation energy that prevents the charging of filters by neutral atoms, allows the neutralisation by ions, and so a quantity of charge of each sign, equal to the figure above, will be removed by the X-ray beam.

Recombination, in which positive and negative ions interact to produce neutral molecules, is the normal fate of ions in air. If this occurred to a significant extent in filters, ions at low concentrations would be more effective in removing charge than those at high, since the rate of recombination is proportional to the square of the concentration (see section on neutralisation of aerosols, in Chapter 9). This is not observed in practice and so it can be concluded that virtually all of the ions remove charge from the filter. However, the rate at which ions are produced in a filter will not be the same as the rate at which they are produced in air, even though 95% of the filter volume may be air. The X-ray photon is much more likely to interact with the material of the fibre than it is with the air, and if it does so it will liberate a high energy photoelectron, which then produces a large number of low energy ions in the interstitial air of the filter. This two-stage process can be accounted for by means of a correction to the free air ionisation, based on modified cavity theory (27), and typical correction factors for common materials are of the order of two.

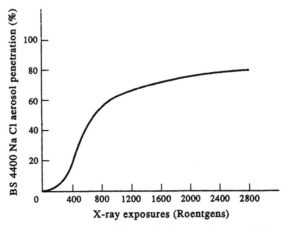

FIG. 5.6 Illustration of loss of charge of electrostatic material as function of integrated radiation dose. The penetration of a standard aerosol through the filter increases with increasing total dose until the filter has lost all of its charge (5). (© Crown Copyright; by permission of Her Majesty's Stationery Office.)

If incremental doses of radiation are given to the filter, and the penetration of an aerosol through the filter is measured as a function of integrated dose, results like those plotted in Fig. 5.6 are obtained. As the integrated dose is increased the penetration increases, eventually reaching a level that does not change with further irradiation. The number of ions required to bring about the neutralisation can be related to the charge that the filter originally held, and this in turn can be related to the charge per unit area of fibre. A problem is that the end-point is difficult to identify with precision. Some specimen results are shown in Table 5.2.

It is clear that in some cases the surface charge density on a fibre exceeds the normal breakdown charge. Air will sustain, over large distances, an electric field of about 3×10^6 volts per metre. Over short distances this field can be exceeded, and a uniformly charged cylinder gives rise to a field of range comparable with its radius (28). However, some of the data in the table exceed even the value that can be held by a fine uniformly charged fibre, which implies that the field is still shorter in range. Since the range of an electric field varies over distances comparable with the periodic distance of its variation, these data indicate that the charge on the fibre has a complicated geometry. It is

TABLE 5.2
Measured charge densities on electrically charged material

| Material | Charge density ($\mu C\ m^{-2}$) | |
	Measured level	Breakdown level for uniformly charged cylinder
Resin wool (sample 1)	850	210
Resin wool (sample 2)	55	210
Split-fibre electret	310	220
Electrostatically spun material	13	350

N.B. The results indicate that material charge can be variable, and that the configuration is complicated. There is no simple relationship observed between charge level and efficiency.

possible that a detailed analysis of the form of variation of curves such as that in Fig. 5.6 would give information on the charge configuration.

Fibre-scanning

An alternative method of charge analysis deals with single fibres removed from the material. Early approaches to this method involved allowing the fibre to settle under gravity in a transverse electric field, and calculating the charge from the angle of descent of the fibre (29). This method, like that using the Faraday ice-pail, measures only the total charge of the fibre; and a fibre that holds positive and negative charge in equal amounts would appear to be neutral.

More sophisticated approaches involve a microscopic study of the filter charge. In one approach (2), the fibre is scanned with a microprobe and electrometer; in another a fibre is suspended vertically, by means of a small weight attached to its end, and a pair of electrodes producing a high field is moved down its length (30). Charge on a segment of fibre will cause it to be attracted to the appropriate electrode as it passes, and the deflection can be related to the charge on the fibre.

These methods give a microscopic picture of the fibre charge, as illustrated in Fig. 5.7, revealing a complicated configuration of charge. A drawback is that the charge on a fibre may be significantly altered by contact with other fibres as it is withdrawn from the material.

Dipolar charge measurement

A final method, applicable only to the materials with corona charge or frozen-in charge, is measurement of the strength of the dipolar charge using a suitable instrument like a capacitive probe (31, 32) or a field

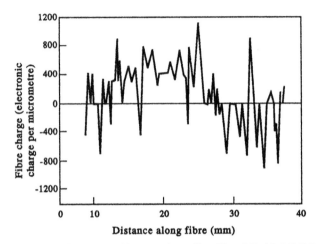

FIG. 5.7   Charge profile observed by scanning a filter fibre (30). (© I.E.E.E., 1986.)

mill, or by thermally stimulated current measurement (33). In the case
of the filters charged as materials the charge measurement can be
carried out on the material itself (33). In the case of the split-fibre
material the charge must be measured on the sheet of fibre-generating
polymer before fibrillation has been carried out (31).

This method applied to a polymer sheet is the least satisfactory, since
although it is undoubtedly capable of giving an accurate measure of the
charge on the polymer at this stage in the manufacture process, it is very
likely that the charge is altered as the material is subjected to the same
sort of processing that results in other types of material developing all
of the charge that they have.

Charge stability and the effect of storage on filter performance

The electrically charged state, being of high energy, is potentially
unstable, and the practical importance of this is that it affects the shelf
life of filters.

Theory of charge loss

A fundamental treatment of the problem of charge loss lies in the realm
of rate theory (34). The electric field within a filter will act on the charge
carriers in such a way as to move them to states of lower energy, but
they resist this because they are held in bound states, surrounded by
energy barriers. If all of the charge carriers were held in states with the

same energy barrier, of height Φ, say, the probability of finding a charge carrier still in its original state after a time t would be,

$$P_\tau(t) = \exp\left(-\frac{t}{\tau}\right) \tag{5.1}$$

where τ is the lifetime of the state, related to the energy by,

$$\tau \sim v \exp\left(-\frac{\Phi}{k_B T}\right) \tag{5.2}$$

and v is a characteristic frequency of the bound charge carrier. At any temperature the charge would decrease exponentially with time, and the time constant quantifying the rate would become rapidly smaller as the temperature increased.

Observed behaviour of filters at elevated temperature

In practice filters do not behave in this way. A filter at any constant elevated temperature suffers a loss of efficiency (an increase in the penetration of an aerosol through it), in a relatively short period of time, after which its behaviour stabilises. If its temperature is raised further, the pattern of behaviour is repeated and the filter again stabilises, but at a higher level of aerosol penetration than before. The temperature dependence of the stable level is illustrated in Fig. 5.8 (35).

This happens because the charge on the filter is held in bound states not with a single binding energy but with a range. At any temperature, charge that is held in relatively shallow levels is lost and that bound more tightly is retained, but the cut off point is altered with the temperature. Behaviour of this kind also occurs immediately after filter material is made (private communication with several manufacturers); the level of performance suffers some deterioration immediately after manufacture but then becomes stable.

Even this picture is incomplete, because the chemical nature of polymers may alter with temperature, affecting the rate of charge loss. Humidity can have an effect on filter performance, and materials with a high moisture regain tend to be more sensitive. Measurements of the penetration of particles of the most penetrating (submicrometre) size through electrostatic filters before and after storage for a period of 42 days at 38°C, 85% RH (36) showed increases of 2–6% from rather high original values of 10–30% shown by the samples tested; and in a separate exercise (37) electrostatic filters stored at 100% RH at

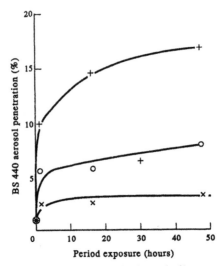

F IG. 5.8 Penetration of standard aerosols through filters exposed to elevated temperatures as a function of time: ×, 90°C; ○, 100°C; +, 110°C (35). (© Crown Copyright; by permission of Her Majesty's Stationery Office.)

ambient temperatures showed no significant increase in penetration. Surface conductivity and the conductivity of interfaces can be affected by adsorbed moisture, but the pattern is not easy to describe (38) and the general picture is clearly very complicated. When materials are humidity sensitive, the effect of elevated temperature and humidity together can be much greater than the effect of either separately.

The general behaviour of filters indicates that there may be some benefit from annealing the charge on the filter, i.e. raising the temperature above normal for a period so that unstable charge is lost. The charge remaining on an annealed filter will be more stable than its original charge, though even without this treatment the charge stability of materials tends to be good, unless they are subjected to extreme conditions.

Results on this effect are relatively few and difficult to generalise, but they do show the pattern suggested by the theory above, and by Fig. 5.8. In particular the rates of loss of efficiency over short and long times are quite different. A situation in which the lifetime of filter charge could be assessed on the basis of a quick test would be ideal, but unfortunately the complexity of the charge loss pattern means that it is difficult to predict the behaviour of material exposed to moderate conditions for a protracted period on the basis of tests carried out in severe conditions for a short period; whereas if all of the charge had the same binding energy this procedure would be easy and reliable.

References

1. HANSEN, N. L. Method for the Manufacture of Smoke Filters or Collector Filters. *British Patent BP* 384052, 1931.
2. WALTON, W. H. The Electrical Characteristics of Resin-Impregnated Filters. *CDE Porton Report No.* 236, 1942.
3. FELTHAM, J. The Hansen filter. *Filtr. Sep.*, 1979, **16**, 370–372.
4. BROWN, R. C. Electrical effects in dust filters. *Proceedings of the 2nd World Filtration Congress*, Filtration Society, London, 1979, 291–301.
5. SMITH, P. A., EAST, G. C., BROWN, R. C. and WAKE, D. Generation of triboelectric charge in textile fibre mixtures and their use as air filters. *J. Electrostatics*, 1988, **21**, 81–98.
6. BROWN, R. C., WAKE, D., BLACKFORD, D. B., EAST, G. C. and SMITH, P. A. The use of electrically charged polymers in air filters. *1st International Conference on the Electrical Optical and Acoustic Properties of Polymers, Canterbury*, 5–7 September 1988.
7. HERSCH, S. P. Resistivity and static behaviour of textile fibres. In *Surface Characteristics of Fibres and Textile Surfaces* (ed. M. J. SHICK), Dekker, New York.
8. SILVERMAN, L., CONNERS, E. W. and ANDERSON, D. M. Mechanical electrostatic charging of fabrics for air filters. *Ind. Eng. Chem.*, 1955, **47**(8), 952–960.
9. TURNHOUT, J. VAN, ALBERS, J. H. M., HOENEVELD, W. J., ADAMSE, J. W. C. and ROSSEN, L. M. VAN. Non-woven electret fibre: a new filtering medium of high efficiency. *Int. Phys. Conf. Ser.*, 1979, **48**, 337–349.
10. TURNHOUT, J. VAN and ALBERS, J. H. M. Electret filters for high-efficiency air cleaning. *Proceedings of 2nd World Filtration Congress, London*, Filtration Society, London, 1979.
11. TURNHOUT, J. VAN, HOENEVELD, W. J., ADANSE, J. W. C. and ROSSEN, L. M. Electret filters for high efficiency and high flow air cleaning. *14th Annual Meeting of the IEEE Industry Applications Society, Cleveland*, 1979.
12. TURNHOUT, J. VAN, ADAMSE, J. W. C. and HOENEVELD, W. J. Electret filters for high efficiency air cleaning. *J. Electrostatics*, 1980, **8**, 369–379.
13. BAUMGARTNER, H. and LOEFFLER, F. Capture of submicrometer aerosol particles by electrostatically charged fibres — an experimental and theoretical study. *3rd International Conference on Electrostatic Precipitation, Abano/Venice, Italy*, 1987.
14. SHIMOKOBE, I., IZUMI, K. and INOUE, M. Electrostatic polarizing of non-woven polypropylene sheets. *Proceedings of the Annual Meeting of the Institute of Electrostatics, Japan, Oct.* 19–20, 1985 (in Japanese).
15. TAYLOR, G. T. Electrically driven jets. *Proc. Roy. Soc.*, 1969, **A313**, 453–475.
16. RARLEIGH. On the instability of a cylinder of viscous liquid under capillary force. *Phil. Mag.*, 1892, **34**, 145–154.
17. SCHMIDT, K. Manufacture and use of felt pads made from extremely fine fibres for filtering purposes. *Melliand Textilber.*, 1980, **61**, 495–497.
18. WEGHMANN, A. Production of electrostatic spun synthetic microfibre nonwovens and applications in filtration. *Proceedings of the 3rd World Filtration Congress*, 1982, Filtration Society, London, 1982.
19. TROUILHET, Y. *Advances in Web Forming*, EDANA, Brussels, 1981.
20. HAYATI, I., BAILEY, A. I. and TADROS, T. F. Mechanism of stable jet formation in electrohydrodynamic atomization. *Nature*, 1986, **319**, 41–43.
21. HAYATI, I., BAILEY, A. I. and TADROS, T. F. Investigations into the mechanisms of electrohydrodynamic spraying of liquids. *J. Coll. Int. Sci.*, 1987, **117**(1), 205–221.
22. WENTE, V. A. Superfine thermoplastic fibres. *Ind. Eng. Chem.*, 1956, **48**, 1342–1346.
23. BROWN, R. C. The behaviour of electrostatic filters made of fibres or sheets arranged parallel to the airflow. *J. Aerosol Sci.*, 1982, **13**, 249–253.
24. BLACKFORD, D. B. and BROWN, R. C. An air filter made from an electret and a conductor. *IEEE Trans. Electr. Insul.*, 1985, **EI-21**, 471–476.
25. LINSHENG, S., BAOJI, C. and YUDE, W. Electret filter used for getting rid of bacteria. *Electrets; 6th Int. Symp. IEEE*, 1988.
26. ATTIX, F. H. and ROESCH, W. C. *Radiation Dosimetry*, Academic Press, New York, 1968.
27. WAKER, A. J. and BROWN, R. C. Application of cavity theory to the discharge of electrostatic dust filters by X-rays. *Appl. Radiat. Isot.*, 1988, **39**(7), 677–684, *Int. J. Radiat. Appl. Instrum., Part A.*

28. HARPER, W. R. *Contact and Frictional Electrification*, OUP, Oxford, 1967.
29. GILLESPIE, T. The role of electric forces in the filtration of aerosols by fibre filters. *J. Coll. Sci.*, 1955, **10**, 299–304.
30. BAUMGARTNER, H., LOEFFLER, F. and UMHAUER, H. Deep bed electret filters. The determination of single fibre charge states and collection efficiencies. *IEEE Trans.*, 1985, **E1-21**, 447–486.
31. FJELD, R. A. and OWENS, T. M. The effect of particle charge on penetration in an electret filter. *IEEE Trans. Ind. Appl.*, 1988, **24**, 725–731.
32. FOORD, T. R. Measurement of the distribution of surface electric charge by use of a capacitive probe. *J. Sci. Instrum.*, 1969, Series 2, **2**, 411–413.
33. ANDO, K., TAKAHASHI, M., TOGASHI, R. and OKUMURA, Y. Properties of electret filter with low pressure drop and high collection efficiency. *Proc. 3rd Int. Aerosol Conf., Kyoto*, Pergamon, 1990.
34. VINEYARD, G. H. Frequency factors and isotope effects in solid state rate processes. *J. Phys. Chem. Solids*, 1957, **3**, 121–127.
35. BROWN, R. C. Modern concepts of air filtration applied to dust respirators. *Ann. Occ. Hyg.*, 1989, **13**(4), 615–644.
36. ONO, M., TONOYA, Y. and TAKENAKA, A. Filtration quality of electret surgical masks after storage in moist air. *Proc. 1985 Ann. Meeting Inst. Electrostatics, Japan*, 158–161 (in Japanese).
37. MOYER, E. S. and STEVENS, G. A. Worst case aerosol testing parameters II. Efficiency dependence of commercial respirator filters on humidity. *Am. Ind. Hyg. Assoc. J.*, 1989, **50**(5), 265–270.
38. ACKLEY, M. W. Degradation of electrostatic filters at elevated temperature and humidity. *3rd World Filtration Congress*, 1982, 176–196, Filtration Society, London, 1982.

CHAPTER 6

Particle Capture by Electric Forces

Introduction

The permanent charges on the fibres of filters described in Chapter 5 increases the efficiency of filters. The efficiency may also be increased by the polarising field of external electrodes. Deliberate charging of the aerosol particles can be used in conjunction with the latter, to increase the efficiency still further.

Capture by permanently charged fibres

When mechanical capture was treated in Chapter 4 both the filter structure and the flow field had to be represented by approximations. The complex configuration of the electric charge likely to occur on fibres produces a further complication when electrostatic effects are included, and approximations to this are necessary if results are to be obtained. A lot of the theory deals with uniformly charged fibres; and the use of dimensionless parameters simplifies much of the working just as it does for mechanical effects.

Figure 5.2 showed the attraction of particles by a uniformly charged fibre, illustrating the drift velocity under the influence of the attractive electric field taking the particle towards the fibre, and the convective velocity of the air tending to carry it past.

Elementary electrostatic theory shows that the electric field at a distance r from a filter fibre carrying a uniform charge Q per unit length acts purely in the radial direction and has a magnitude $\mathbf{E}$ (1);

$$\mathbf{E} = \frac{Q}{2\pi\varepsilon_0 r} \tag{6.1}$$

where ε_0, the permittivity of free space, has the value $10^{-9}/36\pi$ Farads per metre. If the particle has a diameter d_p and an electric charge, q, then its drift velocity towards the fibre will be the product of the force acting on it and its (mechanical) mobility μ, as given by equation 4.34,

$$V_d = \frac{QqCn}{6\pi^2 \varepsilon_o \eta \, d_p r}.$$ (6.2)

Cn, the Cunningham correction factor, defined in equation 4.35, must be included if particles with diameters close to the mean free path of air molecules are considered; but in most equations below it will be understood.

The electrical mobility of a particle, μ_e, is the quotient of the drift velocity, as given by equation 6.2, and the electric field given by equation 6.1, and should not be confused with the mobility, which is the quotient of the drift velocity and the force.

$$\mu_e = \frac{q}{3\pi\eta \, d_p}$$ (6.3)

The quotient of the electrical mobility and the mobility is the particle charge, q. A dimensionless parameter, N_{Qq}, describing the capture efficiency can be obtained by dividing U_d by the convective velocity. Since the drift velocity depends on the distance from the fibre axis, it is evaluated at a fixed distance, corresponding with the fibre surface. Dimensionless numbers are included in the definition of N_{Qq}, for convenience, though they appear as a matter of course in a more rigorous derivation.

$$N_{Qq} = \frac{Qq}{3\pi^2 \varepsilon_o \eta \, d_p \, d_f U}$$ (6.4)

Since the charge on an aerosol particle appears in equation 6.4 with the same power as the fibre charge, it is equally important in determining the capture efficiency by this process. For this reason it is useful to consider the magnitude of electric charge likely to be held by aerosol particles.

Electric charge on aerosol particles

Aerosol charge can take a wide range of values, depending on circumstances. Particles formed as a result of condensation tend to have relatively low levels of charge. Vapour condensing on a nucleus during aerosol formation will neither impart nor remove charge, and so the final charge of the particle will be the same as that of the original nucleus. For the same reason particles produced by the evaporation of a droplet, as in the vibrating orifice aerosol generator (2), may be highly charged, the final particle produced when the evaporation is complete having the same charge as the much larger precursor particle. Particles produced by attrition or fragmentation can be highly charged, and the charge level may depend on the environmental conditions in which the particles are produced. Two theoretical distributions merit special attention, being characteristic of high and low charge levels.

Breakdown charge distribution

A particularly highly charged aerosol is one whose particles have a surface electric field equal to the breakdown field of normal air, which is nominally 3×10^6 volts per metre. The electric field at a distance r from a particle with charge q is,

$$\mathbf{E} = \frac{q}{4\pi\varepsilon_o r^2}.$$
(6.5)

Substituting breakdown field for $\mathbf{E}$ and particle radius for r shows that such a particle, with a diameter d_p micrometres, will hold n_q fundamental charges, where,

$$n_q = 530 \, d_p^2.$$
(6.6)

Equilibrium charge distribution

At the other extreme is the equilibrium or Boltzmann charge distribution, which is the charge distribution that aerosol particles would have if the electrostatic energy of a particle were a degree of freedom in equilibrium with its other degrees of freedom, such as translational motion. It was observed in Chapter 4 that suspended particles hold some of their thermodynamic energy in their translational degrees of freedom, so that they undergo Brownian motion, in which they have an average kinetic energy $\frac{1}{2}k_B T$ associated with random motion in each of the three dimensions, k_B being Boltzmann's constant $(1.381 \times 10^{-23}$ Joules K$^{-1})$ and T the absolute temperature. If electrostatic energy had an equal share of the total energy, then the average particle charge, $\bar{n}_q$, would be given by,

$$\frac{\overline{q^2}}{4\pi\varepsilon_o d_p} = \frac{1}{2}k_B T.$$
(6.7)

Substituting appropriate values of the parameters into equation 6.7 gives, for the mean number of charges on a particle,

Air Filtration

$$\bar{n}_q = 2.4 \, d_p^{1/2}. \tag{6.8}$$

In this situation the particle charge will have a range of values following a Gaussian distribution (3). There is doubt as to whether the equilibrium assumption is strictly correct, but other derivations of the distribution, which do not make such an assumption (4, 5), give values close to that above.

The mean charge given in equation 6.8 ceases to be a useful concept if the particles are so small that they are likely to hold no more than a single charge. The diameter of a particle that holds unit mean charge follows directly from the reasoning leading to equation 6.7, the resulting diameter being 0.11 μm. For smaller particles a reasonable approximation is that a fraction of the particles hold unit charge and that the remainder are neutral.

Miscellaneous charge distributions

Though the two extreme charge levels, equations 6.6 and 6.8, are high and low they are not maximum and minimum. Completely neutral aerosols can be produced in the laboratory by precipitation of all the charged particles in an aerosol; and the normal dielectric breakdown field of the air can be considerably exceeded over distances comparable with particle sizes, the extent of the excess increasing as dimensions become smaller. Over distances comparable with an aerosol particle diameter an extremely high electric field can be maintained, which corresponds to charge levels given approximately by (6),

$$n_q = 1268 \, d_p^{1.7}. \tag{6.9}$$

A number of measurements have been made (7, 8, 9) of the charge level of real industrial and laboratory aerosols, which are found to cover a wide range of values between the above limits. A typical value observed is that of silica at a chemicals factory, which approximates to the form (8),

$$\bar{n}_q = 10.1 \, d_p^{1.2}. \tag{6.10}$$

Capture of charged particles

Returning now to the question of particle capture, we can calculate the drift velocity of aerosol particles for any of the charge distributions, using equation 6.2. For simplicity's sake it will be assumed that the filter fibres hold a charge that would produce the breakdown field at the fibre surface, though just as the normal breakdown field can be exceeded at the surface of small particles, it can be exceeded at the surface of fine fibres. Figure 6.1 shows the drift velocity as a function of diameter for particles holding a charge given by equations 6.6, 6.8 and 6.10, when acted upon by the field at the surface of the charged fibre. Some of the velocities are small compared with the normal convective velocity, and at finite distances from the fibre surface they will be even lower. However a high filtration efficiency can be achieved by a relatively small single fibre collection efficiency, which itself requires only a small drift velocity, and it will be seen below that the practical influence of electric forces on particle capture is considerable.

Capture of neutral particles

An electric field will influence a particle that has no charge of its own because the constituent material of the particle will be polarised by the electric field, which will then attract the induced dipole. The magnitude of the electric dipole is proportional to the electric field strength at the position of the particle and to the particle volume, and it also depends on the dielectric constant of the material making up the particle. Figure 5.2, illustrating this interaction, shows that the component of the dipole closer to the fibre is the component that is opposite in sign, which is attracted to the fibre whatever the sign of its charge. The electric field due to a charged fibre falls off with increasing distance, and so the attracted part of the dipole will be in a slightly higher field than the repelled part. The result of this slight imbalance is a net attractive force on the particle. Exact theory gives the force on a particle caused by polarisation and attraction as,

$$F_r = \frac{\pi \, d_p^3 \varepsilon_o}{4} \left(\frac{D_p - 1}{D_p + 2} \right) \nabla(\mathbf{E}^2). \tag{6.11}$$

The force depends on the gradient of the field rather than on the field itself. A uniform field would produce no force because although it would induce a dipole, the forces acting on the two components of the dipole, being equal and opposite, would cancel out. Substituting equation 6.1 into equation 6.11 gives, for the force due to a uniformly charged fibre,

$$F_r = \frac{Q^2 \, d_p^3}{8\pi\varepsilon_o r^3} \left(\frac{D_p - 1}{D_p + 2} \right). \tag{6.12}$$

The fibre charge appears in this equation as a squared term because

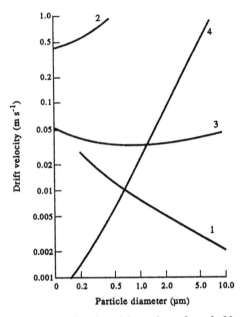

FIG. 6.1   Calculated drift velocity of particles at the surface of a 20 μm diameter fibre with a surface field of 3×10^6 volts per metre: (1) particles with Boltzmann charge distribution;(2) particles with breakdown distribution; (3) particles with distribution $n_q = 10.2 \times d_p^{1.2}$; (4) neutral particles with dielectric constant 4 under the influence of polarisation forces. (N.B. The drift velocities encountered in practice, particularly those of neutral particles, are likely to be lower than those in the situation chosen for illustration.)

it first induces the dipole and then interacts with it, both effects varying linearly with Q. The drift velocity of the particle under the influence of this force is obtained by means of equation 4.34, and the dimensionless parameter appropriate to this form of capture follows from a similar argument to that used for charged particles:

$$N_{Q0} = \frac{Q^2 \, d_p^2}{3\pi^2 \varepsilon_o \eta \, d_f^3 U} \left(\frac{D_p - 1}{D_p + 2} \right). \tag{6.13}$$

The drift velocity for neutral particles of material with a dielectric constant of 4, under the influence of polarisation forces is plotted as a function of particle size in Fig. 6.1 for the situation where the electric field at the fibre surface is equal to the breakdown field of air, as before, and where the filter fibre has a diameter of 20 μm, which is typical of ordinary textile fibres. In some situations the effect of polarisation forces would seem to exceed that of coulomb forces, but this does not mean that neutral particles will be more efficiently captured than charged particles, because polarisation forces act on both neutral and charged particles (the polarisation force is normally neglected when the coulomb force is much larger). In any case, the drift velocity illustrated in the figure is calculated for particles at the fibre surface. For much of its trajectory a particle will be some distance removed, and equations 6.1 and 6.12 indicate that the coulomb forces are much longer in range.

Effect of particle size and filtration velocity

Although the exact way that the single fibre efficiency varies with the dimensionless parameter is not yet clear, it is obvious that as one increases so does the other. This means that for both mechanisms the capture efficiency increases as the filtration velocity decreases, the simple explanation of this being that at low filtration velocities the attractive forces have a longer time in which to act. In the case of polarisation forces the capture is more effective for larger particles, whereas for charged particles the variation with size will depend on the relationship between particle diameter and particle charge. If charge does not vary with diameter, small particles will be more effectively captured. If charge is proportional to diameter the capture efficiency will have no particle size dependence. If the aerosol has a Boltzmann charge distribution, small particles will be more effectively captured, except for very small particles, where those that carry a single charge will be very easily captured, and those that are neutral will not be influenced at all by coulomb forces. For the breakdown charge and for the charge distribution given by equation 6.10, larger particles will be more effectively captured.

Mathematical theory of capture

In general there are three ways of calculating the capture efficiency (10), and these will be summarised below, in such a way that it should be clear which method should be used for which particular charge configuration. The first of these is suitable for the situation where the forces acting are solenoidal (11).

Capture of particles by solenoidal forces

It was shown in Chapter 4, when the capture of particles by gravitational forces was considered, that the problem could be solved exactly because of the solenoidal nature of the force field. The divergence of the field given by equation 6.5 is zero, whereas that of the field in equation 6.12 is non-zero, indicating that coulomb forces are solenoidal, and that polarisation forces are not. In fact these conclusions do not depend on the specific functional form of the force field, which is just the result of the particular configuration of charge on the fibre. They hold for forces of each particular type, irrespective of the form taken by the charge distribution. All coulomb forces are solenoidal, and amenable in principle to treatment by this method. Polarisation forces cannot be treated in this way.

The flow field of the air, which is also solenoidal, can be described by streamlines, the relative density of which is proportional to the local velocity; and the coulomb force field can be described by lines of force, the density of which is proportional to the local magnitude of the force. Inertialess charged particles will follow the streamlines in the former case and the force lines in the latter. In the special case where both flow field and force field act, the two stream functions may be added, and so the particle trajectories can be expressed analytically. A cloud of particles of uniform concentration will, as it approaches a fibre under the influence of both fields, remain uniform in concentration, a property that is critically important since it means that the rate at which a fibre captures particles can be related to the concentration and the velocity, under the influence of the two fields, at the point where capture is deemed to have taken place. The part of the stream function that depends on the force field can be calculated from the following relationships:

$$\frac{\partial \psi_f}{\partial r} = -\mu F_\theta; \quad \frac{\partial \psi_f}{\partial \theta} = \mu r F_r \qquad (6.14)$$

and Fig. 6.2 shows particle trajectories calculated for uniformly charged fibres according to this procedure. There is a stagnation point for particle motion, as indicated; and the limiting trajectory passes through this point, reaching the fibre surface at the rear stagnation point of the air flow, as shown. The value of the total stream function for the particle on the limiting trajectory is,

$$\psi = \frac{Qq}{6\pi \varepsilon_o \eta \, d_p}. \qquad (6.15)$$

Combining equations 6.15 and 4.2 gives the following (exact) expression for single fibre efficiency by coulomb forces under the influence of a uniformly charged fibre,

$$E_{Qq} = \pi N_{Qq}. \qquad (6.16)$$

An important feature of equation 6.16, which is plotted as curve 1 in

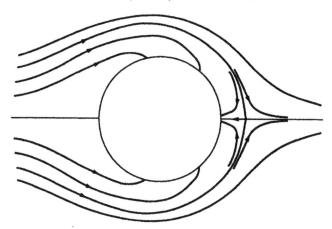

FIG. 6.2 Calculated trajectories of particles captured by a uniformly charged fibre. The limiting trajectory passes through the point where the particle experiences no net force, i.e. the point at the rear of the fibre where the trajectories appear to unite.

Fig. 6.3, is that it does not contain the hydrodynamic factor, nor any other factor depending on the flow field, a feature which is shared by the single fibre capture efficiency for that other purely solenoidal force field, gravity.

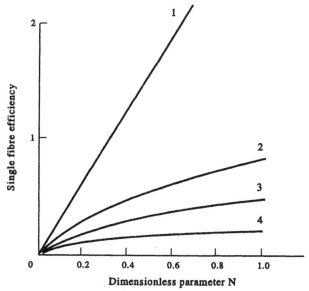

FIG. 6.3 Single fibre efficiency as a function of the relevant dimensionless parameter: (1) uniformly charged-fibre, charged particle; (2) uniformly charged fibre, neutral particle; (3) line dipole charged fibre, neutral particle; (4) $n=2$, line quadrupole charged fibre, neutral particle.

Capture by central forces

A second method of calculation of capture efficiency is appropriate to central forces only (12), and is, therefore, potentially useful for both capture by coulomb forces and capture by polarisation forces, provided that the fibre is uniformly charged. The philosophy behind this method is that if a particle moves as a result of the force and the convective motion of the carrying air, the change in the stream function corresponding to the position occupied by the particle can be related to the force. The incremental change in stream function $d\psi$ can be written quite generally as,

$$d\psi = \frac{\partial \psi}{\partial r}\, dr + \frac{\partial \psi}{\partial \theta}\, d\theta. \tag{6.17}$$

In equation 6.17 the first term is related to the tangential velocity of the air and the second to its radial velocity. If no external force acts, the particle remains on the same streamline and $d\psi$ vanishes. If a central force, $F(r)$, acts, the particle velocity can be described by,

$$V = \mu F_r(r) \tag{6.18}$$

and the motion of the particle can be described by,

$$d\psi = r\mu F_r(r)\, d\theta. \tag{6.19}$$

ψ at distances remote from the particle can be related to single fibre efficiency by equation 4.2, and so the latter can be related to the integral implied in equation 6.19. The shortcoming of the method is that it requires the integration to be carried out along the actual particle trajectory, and this is not known, but must be the subject of a further approximation. It is, however, relatively straightforward to find the limiting trajectory since, in the absence of interception, this is in line with the rear stagnation point. In the case of coulomb forces and a uniformly charged fibre, the right hand side of equation 6.17 can be considerably simplified, and the problem can then be solved exactly, the result being identical to equation 6.16.

Iterative numerical calculations

The third method of calculation, numerical solution of the equations of motion of the particles, is the least elegant, but it is the one method that can be used where all else fails. In general the particles will be subjected to both radial and tangential forces, $F_r(r, \theta)$ and $F_\theta(r, \theta)$ respectively, and so the equations of motion may be written,

$$F_r = \frac{1}{\mu} (U_r - V_r) \tag{6.20}$$

and,

$$F_\theta = \frac{1}{\mu} (U_\theta - V_\theta). \tag{6.21}$$

The end point for the limiting trajectory must be calculated for the particular force, and the trajectory may then be plotted in reverse, ultimately leading to single fibre efficiency. Any numerical method can be used, but the Runge-Kutta method (13) is particularly suitable. This sort of calculation must be distinguished from the calculation of particle trajectories under the influence of particle inertia, for in the latter case the exact conditions of impact, in particular the particle velocity, can not be known in advance, which means that the limiting trajectory cannot be simply identified, and the equation of motion cannot be solved in reverse.

Calculation of single fibre efficiency for polarisation forces

The simplest means of calculating E_{Q0} is the method above (12) for use with central force fields. If the simplest approximation is made, the result is similar to that calculated for the coulomb force, equation 6.16,

$$E_{Q0} = \pi N_{Q0}. \tag{6.22}$$

The reason for the similarity is that the method approximates the force field by a solenoidal field with the same value at the fibre surface. If E_{Q0} is small the approximation is good, but for larger values of E_{Q0} and N_{Q0} the method over-estimates both the force acting on the particle and the single fibre efficiency, which means that the graph of single fibre efficiency against the appropriate dimensionless parameter, plotted as curve 2 in Fig. 6.3, must have the general shape demonstrated there.

If the approximation is made that the capture efficiency is large, then the stream function appropriate for uniform flow can be substituted into equation 6.19 and the resultant expression can be integrated to give (14),

$$E_{Q0} = \left(\frac{3\pi N_{Q0}}{2} \right)^{1/3} \tag{6.23}$$

The eventuality that this expression refers to is unlikely to be realised, but it does indicate the important point that the dependence of E_{Q0} on N_{Q0} becomes weaker as the parameter becomes larger.

The alternative approach (15), fitting a function to the results of numerical calculations of particle trajectories gives, for a Kuwabara flow field with a packing fraction of 0.03, values of N_{Q0} between 0.03 and 0.91,

$$E_{Q0} = 0.84 N_{Q0}^{0.75}. \tag{6.24}$$

Power law fits like this are very useful, frequently giving close approximations to the single fibre efficiency in the form of simple functions, but the function obtained depends on the range of dimensionless parameter considered, and extrapolation outside that range must be avoided.

Particle capture by fibres with non-uniform charge

It was shown in Chapter 5 that the configuration of charge on a filter fibre is often far removed from the simple picture illustrated in Fig. 5.2. The charging process acting during the manufacture of the filter would suggest that fibres in the mixed-fibre material would have a reasonably uniform charge, as may the melt and solvent spun fibres, though this is not confirmed by experiment (16, 17). By similar reasoning the electret filter would be assumed to have a line-dipole configuration, but a considerable amount of fibre–fibre contact takes place after the line-dipole charge has been imparted, and the effect of this on the charge configuration could be considerable. The configuration of charge on the resin wool filter is very complicated indeed, varying not only around the fibre circumference but also along the fibre length.

Electric field caused by fibres with complicated charge configurations

Calculations of electric fields due to these highly complicated configurations of charge are beyond the scope of the relatively simple methods used in this book, though a possible simplifying feature is that electric fields tend to be less variable than the charge causing them because the equation relating the two, Poisson's equation, links the latter with the derivative of the former. In physical terms, the field at any point is caused by the charge at every point.

Detailed calculations will be given below only for two-dimensional charge configurations; the electric charge and field will be assumed, like the airflow, not to vary along the length of the fibre. The electrostatic potential outside a charged fibre can be calculated by the solution of Laplace's equation,

$$\nabla^2 V = 0. \tag{6.25}$$

The approximation that a filter fibre is a circular cylinder and the assumption of a two-dimensional electric charge configuration enables the equation to be written,

$$\frac{\partial^2 V}{\partial r^2} + \frac{1}{r}\frac{\partial V}{\partial r} + \frac{1}{r^2}\frac{\partial^2 V}{\partial \theta^2} = 0. \tag{6.26}$$

If, as treated in detail above, a fibre is uniformly charged, with a charge density of Q per unit length, the electric field outside will be independent of angle and will be given in magnitude by equation 6.1. If the fibre carries a line multipole charge so that the charge varies as the cosine of an integral product of the angular coordinate of the fibre surface, the electric field outside the fibre will be given by the solution of equation 6.26 with the appropriate boundary conditions (1, 18);

$$E_r = \frac{\sigma R^{n+1} \cos n\theta}{\varepsilon_0 (1 + D_f) r^{n+1}} \tag{6.27}$$

$$E_\theta = \frac{\sigma R^{n+1} \sin n\theta}{\varepsilon_0 (1 + D_f) r^{n+1}} \tag{6.28}$$

where $\sigma \cos n\theta$ is the surface density of the fibre charge. The algebraic sum of the charge on a multipole fibre is zero, but the charge of either sign per unit length of fibre, Q, is related to σ by,

$$Q = \sigma\, d_f. \tag{6.29}$$

D_f, the dielectric constant of the material of the fibre, appears in the expression for the field outside a non-uniformly charged fibre, because there is not only an external field but also an internal field, which is affected by the dielectric properties of the fibre. A uniformly charged fibre has no internal field and so D_f does not appear.

The simplest multipole case is that of the line dipole charge configuration, for which the value of n in the above equations is unity. Complicated charge configurations can be built up by adding several multipole distributions together, but only the multipoles themselves will be considered here.

Dimensionless parameters describing capture by multipole charged fibres can be derived by the sort of reasoning used earlier in the chapter, or more rigorously by writing down the equations of motion of the particles. These are very similar to equations 6.4 and 6.13, being, for the charged and neutral particles respectively,

$$N_{\sigma q} = \frac{\sigma q}{3\pi\eta(1+D_f)U\varepsilon_o \, d_p} \tag{6.30}$$

$$N_{\sigma 0} = \frac{2}{3}\left(\frac{D_p-1}{D_p+2}\right)\frac{\sigma^2 \, d_p^2}{\varepsilon_o(1+D_f)^2 \, d_f\eta U}. \tag{6.31}$$

Field lines, the equivalent of streamlines in the airflow pattern, can be calculated for multipole fibres, and the forms of these are illustrated in Fig. 6.4 for the cases of $n=1$ and $n=5$.

Particle capture by multipole charge configurations

The capture of charged and neutral particles by multipole configurations follows from previous work on dipole charge configurations (18, 19). The capture of charged particles can be calculated exactly, since the force field is solenoidal, but a complication is that the efficiency depends upon the orientation of the fibre with respect to the flow, as illustrated in Fig. 6.5. If the charged particle approaches the attractive face of the fibre, the single fibre capture efficiency is a maximum; if it approaches the repulsive face the efficiency is lower and may even be zero. If the distribution of orientations is assumed to be random, an average may be taken; and in the case of a line dipole fibre, the result of the averaging is that, for the range of dimensionless parameters $0.1 < N_{\sigma q} < 10.0$, calculated for a Kuwabara flow field with packing fraction $c=0.05$,

$$E_{\sigma q} = 0.57 N_{\sigma q}^{0.83}. \tag{6.32}$$

For higher orders of charge variation the power dependence on the dimensionless parameter is weaker.

The capture of neutral particles by multipole fibres is, for this type of charge configuration, rather easier to calculate. The two components of electric field can be substituted into equation 6.11, an easy step since E^2 has no angular dependence. The polarisation force is, therefore, central. The use of the simplest approximation will give rise to equations similar to 6.16 and 6.22; and iterative calculation confirms this functional form for small values of E. In the case of multipole configurations the dependence of E or N becomes weaker for increasing

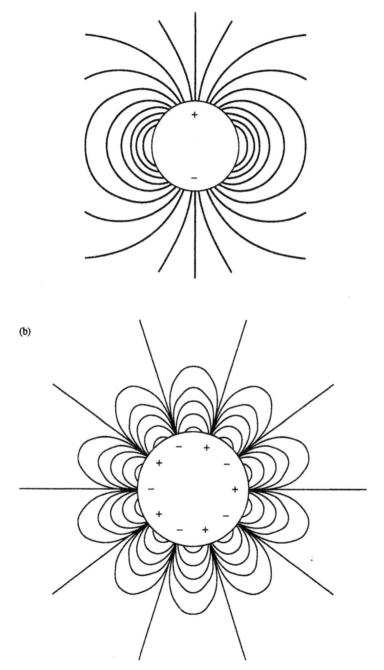

FIG. 6.4 Field lines for particles carrying a charge with angular variability: a line-dipole charged fibre (18). (© Crown Copyright by permission of Her Majesty's Stationery Office); b $n = 5$ charged fibre.

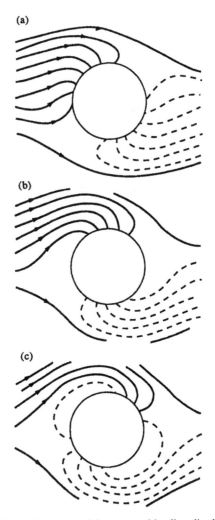

FIG. 6.5   Trajectories of charged particles approaching line-dipole charged fibres, at various orientations. Streamlines that take the form of closed loops will not carry particles (18). (© Crown Copyright by permission of Her Majesty's Stationery Office.)

values of both N and n, as illustrated in Fig. 6.3 (20). The results presented for a dipole configuration obey the following equation, for $1.0 < N_{\sigma 0} < 100$,

$$E_{\sigma 0} = 0.47 N_{\sigma 0}^{0.40}. \tag{6.33}$$

Further study reveals the form of the dependence of the expressions for single fibre efficiency on the hydrodynamic factor. We have seen above that this does not appear explicitly in the expression for capture of charged particles by a uniformly charged fibre, but in fact it appears in all expressions in the following general way (18):

$$E = \frac{Z(N\zeta)}{\zeta}.$$ (6.34)

If E is directly proportional to N, as it is with Coulomb forces due to a uniformly charged fibre, the hydrodynamic factor cancels out, but if E varies according to a fractional power of N, the hydrodynamic factor will appear as a complementary power in the denominator. For example introducing the hydrodynamic factor into equations 6.32 and 6.33 gives,

$$E_{\sigma q} = 0.59 \zeta^{-0.17} N_{\sigma q}^{0.83}.$$ (6.35)

$$E_{\sigma 0} = 0.54 \zeta^{-0.60} N_{\sigma 0}^{0.40}.$$ (6.36)

With knowledge of the dependence of E on the hydrodynamic factor one can generalise calculations based on one type of flow field to situations in which a different type of flow field is observed.

Other workers have obtained the following formulae for the single fibre capture efficiency (21), obtaining functions from the results of calculations, based on the Kuwabara flow field (22):

$$E_{Qq} = \frac{\pi N_{Qq}}{(1 + N_{Qq}^{1/2} c)}$$ (6.37)

$$E_{Q0} = \left(\frac{1-c}{Ku}\right)^{1/8} \frac{\pi N_{Q0}}{(1 + 2\pi N_{Q0}^{1/4})}$$ (6.38)

$$E_{\sigma q} = \left(\frac{1-c}{Ku}\right)^{1/4} \frac{\pi N_{\sigma q}}{(1 + 2.5 N_{\sigma q}^{1/2})}$$ (6.39)

$$E_{\sigma 0} = \left(\frac{1-c}{Ku}\right)^{2/5} \frac{\pi N_{\sigma 0}}{(1 + 2\pi N_{\sigma 0}^{2/3})}$$ (6.40)

The expressions above have a linear dependence on dimensionless parameter when the latter is small, which fundamental theory predicts; and the power at which the hydrodynamic factor appears lies between the two limits that equation 6.34 would predict, for high and low values of the dimensionless parameter.

Combined effect of electrostatic forces with interception and other mechanisms

In the situation where the force due to the electric field is attractive everywhere, the drift velocity, as described above, is directed towards the fibre axis. The convective velocity is usually more or less tangential to the surface and perpendicular to the drift velocity, but the two velocities are parallel at the rear of the fibre, the former diminishing with distance and the latter increasing with it. There exists a stagnation point, indicated in Fig. 6.2, at which a particle experiences no force; and the further from the fibre that this point lies, the more efficient will capture be.

If the particle radius is smaller than the distance of this point from the fibre surface, the physical size of the particle can have no influence on whether or not the particle is captured. If the radius is larger, then the fibre may intercept particles that would not otherwise have been captured. Whether or not interception has any effect in general can be readily understood by plotting the notional radial velocity of a particle that just undergoes contact with the fibre surface. In the model we have used so far, this velocity will equal the radial velocity of the air at a distance of one particle radius from the fibre surface. Two such plots are given, for particles of different size without any electrostatic interaction, in Fig. 6.6, which shows that the velocity is negative (towards the surface) in the front half of the fibre, and positive in the rear half. Particles will only be deposited where the velocity is negative, and so this means that particles may be captured on the front half of the fibre only, which is known to be the case for pure interception.

The addition of an electrostatic force alters the position of the graphs, and the effect of a uniform attractive force is shown in the figure. In the case of the large particle it has increased the angular distance over which capture takes place, and in the case of the small particle it has ensured capture over the entire surface. A repulsive force of the same magnitude would have reduced the capture by an equal amount for the large particle and would have prevented the smaller from being captured at all.

The argument can be made quantitative for the situation where the attractive force is a power law, and the drift velocity towards the fibre is,

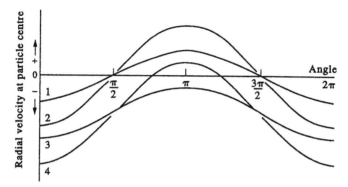

FIG. 6.6   Notional air velocity at one particle radius distance from a fibre surface: (1) pure interception of small particles; (2) pure interception of large particles; (3) pure electrostatic capture of small particules; (4) interception/electrostatic capture of large particles.

$$V_{\mathrm{d}} = \frac{NUR^n}{r^r} \tag{6.41}$$

where N is the (generalised) dimensionless parameter. The radial velocity of the air close to the fibre surface is, according to the simplest approximation, equations 3.10 and 3.34,

$$U_{\mathrm{r}} = \frac{U \delta^2 \cos \theta}{R^2 \zeta} \tag{6.42}$$

where $\delta = r - R$ as before. The particle size at which interception has no effect is given by solving the equation that results from combining equation 6.41 and equation 6.42 for $\cos \theta = 1$, and equating δ with $d_{\mathrm{p}}/2$.

$$\frac{N}{\left(1 + \dfrac{d_{\mathrm{p}}}{d_{\mathrm{f}}}\right)^n} = \frac{d_{\mathrm{p}}^2}{\zeta\, d_{\mathrm{f}}^2} \tag{6.43}$$

Whether or not equation 6.43 can be easily solved depends on the power n, but a simple numerical approach like Newton's method is all that is needed, even for large values of n or for the equation that would result if the complete expression for flow field were used instead of equation 6.42.

When interception is important the value of θ, the angle between the rear stagnation point and the limiting position of particle capture, is given by,

$$\cos \theta = \frac{N\, d_f^2 \zeta}{d_p^2 \left(1 + \dfrac{d_p}{d_f}\right)^n} \tag{6.44}$$

and from this point the method appropriate to central forces can be used to solve the problem.

For the case of a uniformly charged fibre, combination of equations 4.5, 4.7, 6.16 and 6.43 shows that the single fibre efficiency at which interception can be neglected is related to the single fibre efficiency that would be observed in the case of pure interception, by (19, 20),

$$E_{Qq}(\min) = \pi E_R \tag{6.45}$$

and the general expression for single fibre efficiency, where interception cannot be neglected, is,

$$E_{\min} \le \pi E_R. \tag{6.46}$$

In the case of complicated force laws like that between a line-dipole charged fibre and a charged particle, the effect of interception depends on the orientation (18). Moreover particles are not necessarily captured at points where the radial velocity is negative because, as indicated in Fig. 6.5, in some instances the "trajectories" are closed loops, not encountering a particle source.

Calculations have been made of the combined effects of electric forces and gravity (23) using analytical methods, but it is difficult to write down generalised formulae and it seems that for situations like these, each specific example needs to be worked through. The reader is referred to the original work for the results.

Capture by image forces

The final electrostatic mechanism that can affect filtration efficiency results from the force between an electrically charged particle and a neutral fibre. Although this type of force is usually referred to as an image force, and the force between a charged fibre and a neutral particle is called a polarisation force, the mechanism in the two cases is the same. It is only the site of the charge that differs.

The theory of electrical images (1) seeks to replace the charge distribution developed by a conducting or dielectric body in response to an external charge, by a geometrically simpler hypothetical "image" charge distribution. The image charge must produce the same external field as the real induced charge or polarisation charge, and this condition is satisfied if it gives rise to the physically correct boundary conditions. An electrical image charge in a conductor can equal in magnitude the charge producing the image; that in a dielectric is smaller.

Electrical images in a planar body closely resemble simple optical reflection images. Those in spherical bodies are more complicated but still amenable to simple analytical treatment. Those in cylinders can be described analytically only in terms of special functions. However, provided that the capture efficiency is not too high, only the situation where the inducing charge is close to the surface needs to be considered, and in this situation the curvature of the surface can be neglected and the surface approximated by a plane. A further approximation is replacement of the charge on the particle by a point charge, which is sound unless the particle is very large. With these approximations the force acting on the charged particle as a result of its image in the fibre, is,

$$F_r = \left(\frac{D_f - 1}{D_f + 1}\right) \frac{q^2}{16\pi\varepsilon_o \, \delta^2}; \quad \text{where} \quad \delta = r - R. \tag{6.47}$$

The stream function close to the fibre surface can be expressed, to the same level of approximation, by the simple form given in equation 3.34,

$$\psi = \frac{2U \sin \theta \, \delta^2}{\zeta \, d_f}. \tag{6.48}$$

Equation 6.47 can be substituted into equation 6.19 which is valid in this situation because the image force is a central force; and the distance δ can be eliminated using equation 6.48. The integration is then straightforward, and the result is the following expression for the single fibre efficiency,

$$E_{0q} = \frac{2}{\zeta^{1/2}} N_{0q}^{1/2} \tag{6.49}$$

where N_{0q}, the dimensionless parameter describing particle capture by image forces, is given by,

$$N_{0q} = \left(\frac{D_f - 1}{D_f + 1}\right) \frac{q^2}{12\pi^2 \eta U \varepsilon_o \, d_p \, d_f^2}. \tag{6.50}$$

Figure 6.7 illustrates the behaviour of a particle moving under the influence of image forces.

Measurements of capture by image charges have been carried out

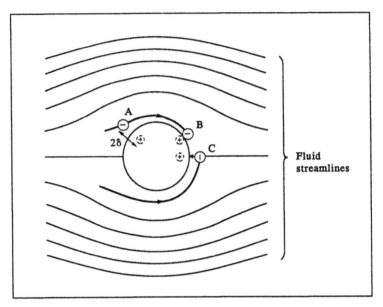

FIG. 6.7 Capture of particles by image forces: A, particle attracted by its image in the fibre surface; B, captured particle; C, particle on limiting trajectory (the distance from the fibre surface of the point at which the particle experiences no net force is larger than the particle radius, and so interception does not affect the capture efficiency of this particle).

using particles of micrometre and submicrometre size (24) and neutralised glass fibre filters, some of which had fibres that were silver coated. In general the observed single fibre efficiencies were higher than predicted by equation 6.49, but the functional dependence on N_{0q} was similar, and the following expression gave a good fit to the experimental results,

$$E_{0q} = 1.5 N_{0q}^{1/2}.\qquad(6.51)$$

Experiments by other authors (25) on a glass fibre filter and with aerosol particles of comparable size gave results fitted by,

$$E_{0q} = 2.3 N_{0q}^{1/2}.\qquad(6.52)$$

Combined effects of image forces and interception

The combined effect of interception and image forces can be calculated easily because the simplified expressions for both the flow velocity and the force acting on the particles close to the fibres depend on a power of the distance from the fibre surface. The point, in line with the rear stagnation point, where these two balance, can be found from equations 6.47 and 6.48; and interception can be neglected provided that this is further from the surface than one particle radius, i.e. provided,

$$N_{0q} > \frac{N_R^4}{\zeta} \tag{6.53}$$

When this is not the case the expression for the single fibre efficiency as a result of the combined effects is,

$$E_{R0q} = \left[\frac{\sin^2 \theta N_R^4}{\zeta^2} + \frac{2N_{0q}(1 + \cos \theta)}{\zeta} \right]^{1/2} \tag{6.54}$$

where θ is the angle, measured from the rear stagnation point, at which the particle on the limiting trajectory touches the surface of the fibre. The value taken by θ is that which results in the bracketed expression being a maximum (otherwise it would not be the limiting trajectory), which is,

$$\cos \theta = \frac{N_{0q}\zeta}{N_R^4}. \tag{6.55}$$

Experimental observation of electrostatic capture

The difficulties of reconciling theory and experiment for the capture of particles by mechanical effects were apparent in Chapter 4, where it was shown that the difference between the simplified and idealised structure of theoretical models and the highly complicated structure of real filters made discrepancies inevitable. However, general trends were clear and, in particular, the dependence of capture efficiency on certain dimensionless parameters could be demonstrated.

Experiments on electrically charged filters have these complications and further ones besides. Although electric charge can be eliminated when mechanical effects are studied, it is not possible to eliminate mechanical effects in a study of the electrical behaviour of filters. Moreover, there is no completely reliable method of measuring the configuration of charge on the fibres of a filter *in situ*, and some assumption must be made about this before the results of experiments can be interpreted. The sections above have shown that the effect of charge configuration on the single fibre efficiency itself is considerable; and experimental results and their interpretation must be qualified in this way.

Just as the understanding of mechanical processes in filtration has benefited from the use of monodisperse aerosols, the study of electrical forces in filters is advanced considerably by the use of aerosols consisting of particles that are not only all of the same size, but also all in the same charge state; and this introduces an experimental problem. It is not easy to control charge exactly, though it is possible repeatedly to induce charges on particles produced from a vibrating orifice generator (26), and selection can be carried out with submicrometre particles which, as normally produced, have either a single charge or none. The singly charged particles can be precipitated out, to leave a completely neutral aerosol; or the particles may be classified electrostatically, so that a certain fraction with a prescribed electrical mobility are selected, giving particles that all have a single charge of the same sign.

An indication of the importance of electric forces, is however, easy to obtain. If the efficiency of a filter in its normal charged condition is compared with its efficiency after the entire charge has been removed by a massive dose of ionizing radiation, a considerable difference can be observed. The results of such an experiment (27) have been shown in Fig. 5.1.

More detailed measurements have been made by a number of observers, mostly with submicrometre aerosols on filters in their charged state, and the results of some of these are illustrated in Figs 6.8–6.11. It is not easy to compare the results obtained by different workers on different occasions, because not only do filter material types vary, but also thickness, area weight, pressure drop and filtration velocity; and results expressed as a penetration may refer to conditions in which these parameters take any of a wide range of values. However a number of different results are given below, in the form of a quality factor plotted against particle diameter.

Results obtained by one set of workers (16, 28, 29, 30, 31) are illustrated in Figs 6.8–6.10, and it is clear from these that in every instance more than one filtration mechanism is operating. In Fig. 6.8 measurements obtained with a neutral aerosol and neutral filter material are given as baseline results. The filtration efficiency has a minimum, caused by the combination of interception, which is more effective at large particle size, and diffusional deposition, which is more effective for smaller particles. Results obtained with the same filter in the charged state show considerable difference at the larger particle sizes, because of polarisation forces, which are larger at larger particle

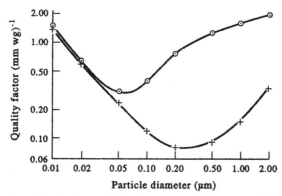

FIG. 6.8   Capture of neutral aerosols by filters, expressed in terms of quality factor (equation 1.8): ⊙, filter charged; +, filter neutral (29). This figure, and Figs 6.9–6.12, illustrate that quality factor is a function of particle size and charge, and of the air velocity during filtration. (By permission of Springer-Verlag, Heidelberg.)

sizes; but at small sizes diffusion is still the dominant process, and the general result has been to shift the size at which filtration efficiency is a minimum to a smaller value.

Figure 6.9 shows the size dependence of filtration of neutral particles by charged fibres over a range of filtration velocities. It is clear from the graphs that the two competing mechanisms are both more efficient at lower filtration velocities; and the two mechanisms are identified with diffusional deposition and polarisation capture as before. The position of the minimum in the efficiency curve is not greatly altered by changes in velocity of the order shown, and is similar to the most penetrating particle size observed with charged filters and neutral particles by other workers (32).

Figure 6.10 gives comparable results for charged particles. The original data, in which filter efficiency rather than quality factor was plotted, showed more clearly the difference in the filtration efficiency between charged and neutral particles, but although the data in Fig. 6.10 still indicate this, the method of presentation makes it less obvious. The additional mechanism operating, coulomb attraction between the fibres and the particles, makes a difference at all particle sizes. Again the position of the size at which filtration efficiency is minimal is not critically dependent upon velocity; but is observed at a rather smaller particle size than that observed by other workers.

One would not expect perfect agreement among the results of all workers (21, 33, 34, 35, 36, 37) because of unavoidable experimental differences, but similar general features are clear. Figure 6.11, reported by other workers (36), shows the behaviour of singly charged particles with charges of either sign, as a function of particle size. The capture of small particles is more effective than that of large over the range of

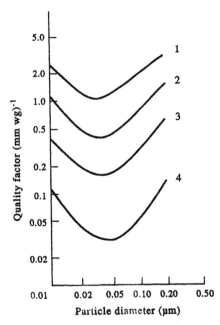

FIG. 6.9 Capture efficiency of neutral aerosols by charged filters: (1) $U = 0.055\ ms^{-1}$; (2) $U = 0.1\ ms^{-1}$; (3) $U = 0.2\ ms^{-1}$; (4) $U = 0.5\ ms^{-1}$ (29). (By permission of Springer-Verlag, Heidelberg.)

parameters used, indicating that a single mechanism is dominant, which is likely to be capture by coulomb forces.

The same authors calculate that for a Boltzmann distribution, the penetration curve should have two local maxima and therefore a minimum in between; and their experimental values cover the range where a minimum would be expected (33, 37). The existence of such a minimum is not a feature of all measurements, but since it depends critically on a particular aerosol charge distribution, absolute repeatability would be difficult to achieve.

Figure 6.12 shows a comparison of quality factors for the four cases in which aerosol and filter are charged and neutral (38). The most obvious feature of the results is that the size of the most penetrating particle differs according to the charged state of both the fibre and the particles. If the fibre is charged the particle size is larger for the singly charged particles than it is for neutral, confirming the observations of earlier workers (39), since the high electrical mobility of small particles increases the capture efficiency, whereas the effect of a single charge on large particles is much less marked. In the case of a neutral filter, no significant difference is observed in the data shown, though in different experimental conditions there is a possibility that the possession of a

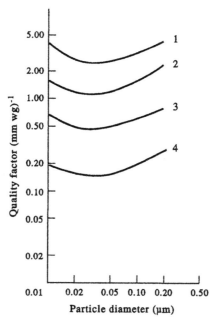

FIG. 6.10 Capture efficiency of charged aerosols by charged filters at a range of velocities: (1) $U=0.055$ ms^{-1}; (2) $U=0.1$ ms^{-1}; (3) $U=0.2$ ms^{-1}; (4) $U=0.5$ ms^{-1} (29). (By permission of Springer-Verlag, Heidelberg.)

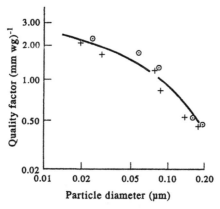

FIG. 6.11 Capture efficiency of charged aerosols by charged filters: $\odot$, negative particles; $+$, positive particles (36). (With permission from Springer-Verlag, Heidelberg.)

charge could increase the size of the most penetrating particle, because the augmentation of filtration efficiency by image forces is greater at small particle size.

In the case of neutral particles, the charge on the fibre will reduce the

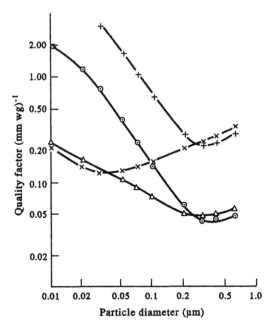

FIG. 6.12 Capture efficiency of aerosols by filters: △ particles neutral, fibres neutral; ⊙ particles singly charged, fibres neutral; × particles neutral, fibres charged; + particles singly charged, fibres charged (38). (© Crown Copyright; by permission of Her Majesty's Stationery Office.)

size of the most penetrating particle, because polarisation forces are greater on large particles, and this is supported by some data from other observers (40, 41).

Capture efficiency compared with dimensionless capture parameters

A further interesting exercise is to plot experimental results against an appropriate dimensionless parameter, though, since the dimensionless parameters appropriate to any charge configuration are similar, dependence on a particular parameter indicates only that a particular mechanism is operating (i.e. coulomb capture or polarisation capture). The functional form of that dependence must be studied if information about the charge configuration is required.

Results from two sets of workers (36, 42) are plotted in the form of single fibre efficiency against the dimensionless parameter describing capture, in Fig. 6.13, and both illustrate that dependence is relatively weak. Uncertainties or errors in the filter structure or charge level would translate the curves but not alter their shape.

Other workers (43) have charged, by corona, filters of identical structure, measuring the filtration efficiency as a function of the charge

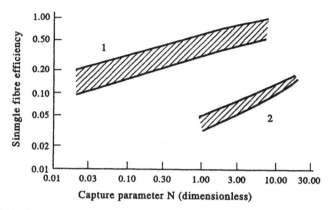

FIG. 6.13 Measured dependence of single fibre efficiency on dimensionless parameter; (1) reference 36 (with permission from Springer-Verlag, Heidelberg); (2) reference 42 (© 1988, I.E.E.E.)

measured on the entire filter. It is difficult to relate such results to those above, but dependence on the dimensionless parameter is similarly weak. The single fibre efficiency, calculated from the experimental data, varies by a factor of only 1.6 whilst the charge varies by 2.8.

If a fibre were uniformly charged and acting upon charged particles, the range of force acting would be as long as is possible with the particular geometry, and the capture efficiency would be proportional to the parameter. If, however, the action is on uncharged particles and if the configuration of charge is more spatially variable, the dependence of capture efficiency on the parameter is weaker as shown in Fig. 6.3. The results in Fig. 6.13 therefore indicate that the charge is spatially variable, which is consistent with the quantification of charge carried out by scanning single fibres (16), and by irradiating whole filters (44).

Functional form of single-fibre efficiency by electrostatic effects and diffusion

Experimental measurements over a wide range of experimental parameters have been made (32) and a function describing the single fibre capture efficiency, containing the three dimensionless parameters appropriate to capture by diffusion, coulomb forces and polarisation forces, has been fitted to a large number of experimental results.

$$E = 1.07 Pe^{-2/3} + 0.06 N_{Q0}^{2/5}$$
$$+ 0.067 N_{Qq}^{3/4} - 0.017 (N_{Q0} N_{Qq})^{1/2} \qquad (6.56)$$

The coefficient of the Peclet number, describing diffusion, is lower than results cited in Chapter 4 by about a factor of 2. The coefficients of the other parameters are considerably lower than the working above would suggest, but it appears that the calculations have been carried out on the assumption of a unipolar charge for the fibre whereas the type of filter used in the experiments may be better described by a line dipole charge. In addition a value has had to be assumed for the charge held by the filter, and as the authors say, this value cannot be verified. The most important feature of the work is that the values of the exponents are completely consistent with the theory given above, though they have not been imposed on the data, but fitted to the experimental results by plotting the data on log–log plots. The last term in the expression, because of its negative coefficient, will limit the range of validity of the expression.

Augmentation of filtration efficiency by external electric fields

Just as the polarisation charge of a particle can increase filtration efficiency, so can the polarisation charge of a filter fibre when it is placed in an external electric field. A cylindrical dielectric fibre placed in a uniform electric field, $\mathbf{E}$ develops a line dipole with a moment, m_p, per unit length where,

$$m_p = 2\pi\varepsilon_o \mathbf{E} R^2 \left(\frac{D_f - 1}{D_f + 1}\right). \tag{6.57}$$

Any particles passing through the filter will feel the effects of both the external field and the polarisation field caused by the induced dipole. A particle with coordinates (r, θ) relative to the fibre axis will be subjected to the following field components, when the electric field is acting along the direction $\theta = 0$,

$$\mathbf{E}_r = \left[\left(\frac{D_f - 1}{D_f + 1}\right)\left(\frac{R}{r}\right)^2 + 1\right] \mathbf{E} \cos\theta \tag{6.58}$$

$$\mathbf{E}_\theta = \left[\left(\frac{D_f - 1}{D_f + 1}\right)\left(\frac{R}{r}\right)^2 - 1\right] \mathbf{E} \sin\theta. \tag{6.59}$$

Capture of charged particles

Charged particle trajectories can be calculated analytically (11); and study of the equations, or use of arguments similar to those at the beginning of the chapter, results in the emergence of a dimensionless parameter for this mode of capture.

$$N_{pq} = \frac{Eq}{3\pi\eta \, d_p U} \tag{6.60}$$

The capture efficiency depends, as it does in the case of capture by permanently charged fibres, on the orientation of the dipole with respect to the direction of flow. The simplest means of applying an external field is by means of grid electrodes on the leading and trailing faces of the filter, and in this situation the electric field is parallel to the flow field, and the single fibre efficiency for capture of charged particles can be calculated to be,

$$E_{pq} = N_{pq} \left[\frac{\left(\dfrac{D_f - 1}{D_f + 1} \right) + 1}{N_{pq} + 1} \right]. \tag{6.61}$$

Capture of neutral particles

The mechanisms of capture of a charged particle by a permanently charged line dipole fibre and by a fibre in an external electric field have a lot in common, but the mechanism of capture of a neutral particle by a polarised fibre is substantially different from that of capture by permanently charged fibres. In the latter case the same electric field both produces the induced dipole and attracts it, but in the former case it is the external field that polarises the particle, and the (much smaller) field of the polarised fibre that attracts it.

Acting alone the external field would induce a dipole with a moment that is parallel to that of the fibre, and with the same strength whatever the position of the particle. The external field will, being uniform, not influence the motion of the dipole that it has induced (*vid sup*). The two aligned dipoles, that of the fibre and that of the particle, will attract if they are in line, but not if they are parallel, as indicated in Fig. 6.14. The complicated pattern of force around the polarised fibre (45) contrasts with the simple inverse fifth power law that exists around a simple permanently polarised fibre. A dimensionless parameter describing the interaction between a polarised fibre and a neutral particle can be derived.

$$N_{p0} = \frac{2}{3} \left(\frac{D_p - 1}{D_p + 2} \right) \left(\frac{D_f - 1}{D_f + 1} \right) \frac{d_p^2 \varepsilon_o \mathbf{E}^2}{d_f \eta U} \tag{6.62}$$

The single fibre efficiency for capture is a complicated function

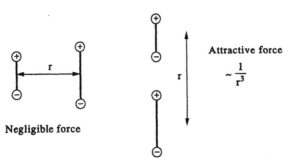

Fig. 6.14   Forces between a line dipole and a point dipole, both induced by an external electric field, aligned parallel and in line.

matching the complicated force law, but for low values of the dimensionless parameter the relationship is linear and independent of the form of the flow field (*vid sup*).

$$E_{p0} = \frac{N_{p0}}{2} \qquad (6.63)$$

As N_{p0} increases, E_{p0} increases progressively more weakly, just as for the corresponding parameters with permanently charged fibres. Other calculations (46) predict that E_{pq} will also vary more weakly with N_{pq} as the latter increases.

Further theoretical work (47), using the Kuwabara model (22), has led to a slightly different result from equation 6.61:

$$E_{pq} = \frac{N_{pq}\left[\left(\dfrac{D_f-1}{D_f+1}\right)+1\right]}{N_{pq}\left[c\left(\dfrac{D_f-1}{D_f+1}\right)+1\right]+1} \qquad (6.64)$$

where c is the packing fraction. As part of the same exercise numerical calculations were carried out for neutral particles, resulting in values similar to those calculated above but lower by about 20%. Adding the refinement of aerodynamic slip reduced the discrepancy to about 15%. Further theoretical calculations have included the effects of gravity (23), particle inertia (48, 49), diffusion (50) and aerodynamic slip (51).

Direct observation of particle trajectories

Visual confirmation of the predicted results has been made using isolated fibres placed between electrodes and then exposed to aerosols (52, 53). The Reynolds number for flow had been kept low enough for the Lamb field to be a good description of the flow pattern, and particles were injected into the flow and traced by intermittent illumination. In some instances the fibres had a unipolar charge as well as a polarisation charge. The particles were also charged and this meant that a stream function could be written down describing the behaviour analytically, and both the unipolar charge and the fibre's polarisation charge could be included in the theory.

$$\frac{\psi}{U} = \frac{R}{[2.0022 - \ln(Re)]} \left[\frac{R}{r} - \frac{r}{R} + 2\frac{r}{R} \ln\left(\frac{r}{R}\right) \right] \sin\theta$$

$$+ N_{pq} \left[\left(\frac{D_f - 1}{D_f + 1}\right) \frac{R^2}{r^2} + 1 \right] r \sin\theta + N_{Qq} R\theta \qquad (6.65)$$

Figure 6.15 illustrates that agreement between the calculated trajectories and the experimentally observed ones was good.

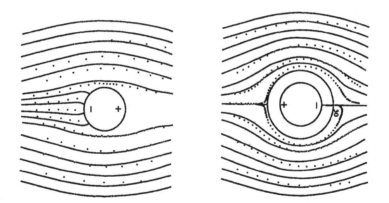

FIG. 6.15 Calculated and observed trajectories of charged particles under the influence of a fibre in an external electric field: a, particles approaching the attractive pole; b, particles approaching the repulsive pole (52). (With permission from Springer-Verlag, Heidelberg.)

Experimental measurements on model filters

Experimental tests of equation 6.63 have been carried out on model filters consisting of layers of wires or polymer fibres in either staggered or fan-model geometry (54). The aerosols used were uncharged monodisperse aerosols of DBP and selenium with diameters of 1.32 and 4.2 μm. A large number of measurements were carried out in a variety of experimental conditions, and the results were fitted to a function of the form:

$$E_{p0} = \alpha N_{p0}^{\gamma} \tag{6.66}$$

the parameters α and γ being constants. The dielectric constant of the metal fibres was infinity, and that of the aerosols was assumed to be similar; but whatever assumptions were made, the power law in equation 6.66 would not be affected. The values fitted, for a range of $0.0005 < N_{p0} < 0.5$, fall on a straight line with $\alpha = 1$ and $\gamma = \frac{2}{3}$, which is a rather weaker variation than theory predicts.

Effect of single fibre position and composition

Other workers made measurements (55, 56) with single fibres, but using particles that were effectively neutral, that is to say that their charge level was sufficiently low for the polarisation forces far to outweigh the coulomb forces in intensity. They studied capture, but not particle trajectories, using both metal fibres and polymer fibres; and found that for a wide range of parameters the experimental values of the single fibre capture efficiency fell close to a single line that could be roughly described by a power law in the dimensionless capture parameter. There did not appear to be any significant difference between the capture efficiency of metal fibres and polymer fibres, indicating that the latter were effectively conductors in this situation.

The capture efficiency of a single conducting non-isolated fibre increased as it was moved from its midway position to a position closer to one of the planar electrodes than to the other. When midway the fibre would carry purely a line dipolar induced charge; but there would be a unipolar element in addition if the fibre were nearer to one electrode, the magnitude of which would vary with the displacement of the fibre from the midpoint. The difference in the capture efficiency of the fibre in the two positions was up to an order of magnitude, indicating the much higher strength of interaction resulting from unipolar than from dipolar charge.

The above effect can easily be calculated or observed for a single fibre, but it is not easy to generalise the effect to a filter composed of many fibres filling the space between the electrodes. Most of the fibres will be closer to one electrode than to the other, and the closest of all will, if they are conducting and earthed, become charged in this way; but the effect of their charge will be to screen the electrodes and to reduce both the electric field and the filtration efficiency in the central regions of the filter. It is not certain which effect will dominate.

In particular, one would expect that in the case of a filter made of conducting fibres there would be a screening distance (which would depend on the packing fraction) beyond which the electrode charge would be screened out completely. The efficiency of filters would not be expected to vary with thickness for thicknesses greater than about twice this distance. An isolated dielectric fibre will be more highly polarised if it is near to an electrode because it will be closer to its image dipole, but this will not occur if the entire space between the electrodes is filled with fibres. In this case, the fibres will be uniformly polarised, and the efficiency of the filter should vary with its thickness in the normal way.

Measurements on real filters in an external electric field

Experimental measurements have been carried out (57) using approximately monodisperse aerosols and filters made from both conducting and dielectric fibres. Considerable augmentation of filtration efficiency was observed, but the two types of filter differed, in that the increase in single-fibre efficiency of the filter made from conducting fibres was greater at low packing fractions than at high, where as that of the dielectric fibre filter did not vary. Furthermore, the aerosol penetration through the dielectric filter decreased exponentially with its thickness, indicating uniformity of conditions within the filter.

Other authors (58) also observed this exponential variation of aerosol penetration. In addition the efficiency was observed to increase as both the fibre diameter and the filtration velocity were decreased; but for a 1.0 μm diameter aerosol, there was no difference in filter penetration between particles carrying on average 13 elementary charges, and particles with approximately a Boltzmann charge distribution.

The effect of an external electric field on the penetration of particles of sizes 0.01–2.0 μm has been shown (59) to be greatest at the region of maximum penetration, reducing this by an order of magnitude. The effect at other particle sizes is less, which reduces the size dependence of aerosol penetration.

A very significant practical point is the rate of clogging which, for the same amount of deposited dust, is several times more severe in the case of the filter without the field. The effect has been ascribed to the more uniform deposition of dust around the entire fibre surface in the case of the electrically enhanced filter, though the influence of the field on the microscopic structure of the deposit also appears to be important. The effect is also observed by other workers (60), and is consistent with theoretical predictions (61).

Augmentation of filtration efficiency by charging of particles

The efficiency of filters with external fields can be further increased by deliberate charging of the particles; and a practical use has been made of this effect for the upgrading of HEPA filters for clean room applications (62, 63, 64). No detailed theory, nor measurements of charge were made in this study, and the low penetration of aerosols through the filter (0.01%) without electrical augmentation indicated the action of other significant capture mechanisms; but penetrations of particles larger than 0.1 μm was observed to be as low as 0.0005% when charging was carried out.

The influence of pre-charging of particles subsequently to be passed into an uncharged filter has been studied experimentally (64) and theoretically (65) but with respect to fabric filters. Not only was the efficiency found to be enhanced, but also the effect of loading a pressure drop was reduced, probably as a result of the formation of a porous cake because of particle–particle repulsion. The effect of particle pre-charging also improves the performance of filters of moderate efficiency with an applied electric field (51, 66).

References

1. FEWKES, J. H. and YARWOOD, J. *Electricity and Magnetism*, University Tutorial Press, London, 1956.
2. BERGLUND, R. N. and LIU, B. Y. H. Generation of monodisperse aerosol standards. *Env. Sci. Tech.*, 1973, 7, 147–153.
3. KEEFE, D., NOLAN, P. J. and RICH, T. A. Charge equilibrium in aerosols according to the Boltzmann law. *Proc. Roy. Irish Acad.*, 1959, **60A**, 27–45.
4. FUCHS, N. A. On the stationary charge distribution on aerosol particles in a bipolar ionic atmosphere. *Geofis. Pura. Appl.*, 1963, **56**, 185–193.
5. BROWN, R. C. Theory of interaction between aerosol particles and an ionized gas in the low concentration limit. *J. Aerosol Sci.*, 1991, **22**(3), 313–325.
6. HARPER, W. R. *Contact and Frictional Electrification*, OUP, Oxford, 1967.
7. VINCENT, J. H., JOHNSTON, A. M., JONES, A. D. and McLACHLIN, C. Q. *IOM Report No. TM/83/15*, 1983.
8. JOHNSTON, A. M., VINCENT, J. H. and JONES, A. D. Measurement of electric charge for workplace aerosols. *Ann. Occ. Hyg.*, 1985, **29**(2), 271–284.
9. JOHNSTON, A. M., VINCENT, J. H. and JONES, A. D. Electrical charge characteristics of dry aerosols produced by a number of laboratory mechanical dispensers. *Aerosol Sci. Tech.*, 1987, **6**(2), 115–127.
10. PICH, J. Analytical method for the calculation of deposition of particles from flowing and central force fields in aerosol filters. *Env. Sci. Technol.*, 1977, **11**, 606–612.
11. DUCHIN, S. S. and DERYAGUIN, B. V. Towards a method of calculating the precipitation of disperse particles from a flow interrupted by obstacles. *Kolloid Zh.*, 1958, **20**, 326–328 (in Russian); HSE Translation 7221 (1978).
12. NATANSON, G. L. Deposition of aerosol particles by electrostatic attraction upon a cylinder around which they are flowing. *Dokl. Akad. Nauk, USSR*, 1957, **112**, 696–699 (in Russian); HSE Translation 7222 (1977).
13. AYRES, F. *Theory and Problems of Differential Equations*, Schaum, New York, 1953.
14. KRAEMER, H. F. and JOHNSTONE, H. F. Collection of aerosol particles in presence of electrostatic fields. *Ind. Eng. Chem.*, 1955, **47**(12), 2426–2434.

15. STENHOUSE, J. I. T. The influence of electrostatic forces in fibrous filtration. *Filtr. Sep.*, 1974, 25–26.

16. BAUMGARTNER, H., LOEFFLER, F. and UMHAUER, H. Deep-bed electret filters — The determination of single fibre charge state and collection efficiency. *IEEE Trans.*, 1985, E-21(3), 477–480.

17. LATHRACHE, R. Private communication.

18. BROWN, R. C. The capture of dust particles in filters by line-dipole charged fibres. *J. Aerosol Sci.*, 1981, 12(4), 349–356.

19. PICH, J., EMI, H. and KANAOKA, C. Coulombic mechanism in electret filters. *J. Aerosol Sci.*, 1987, 18(1), 29–35.

20. THORPE, A. and BROWN, R. C. A study of the electric charge in filters using monodisperse test aerosols. *5th Aerosol Society Conference Proceedings*, 1991.

21. LATHRACHE, R. and FISSAN, H. Grundlegende Untersuchungen zum Abscheideverhalten der Elektret-Filter Teil 1: Bestimmung der Abscheidegrade. *Staub Reinhalt. Luft.*, 1989, 49, 309–314.

22. KUWABARA, S. The forces experienced by randomly distributed parallel cylinders or spheres in viscous flow at small Reynolds numbers. *J. Phys. Soc.*, *Japan*, 1959, 14, 527–532.

23. NIELSEN, K. A. Collection of inertialess particles on circular cylinders with electrical forces and gravitation. *J. Coll. Int. Sci.*, 1978, 64, 131–141.

24. LUNDGREN, D. A. and WHITBY, K. T. Effect of particle electrostatic charge on filtration by fibrous filters I & EC. *Process Design and Development*, 1965, 4, 345–349.

25. YOSHIOKA, N., EMI, H., HATTORI, M. and TAMORI, I. Effect of electrostatic force on the filtration efficiency of aerosols. *Kagaku Kogaku*, 1968, 32, 815–820 (in Japanese); HSE Translation 7452 (1977).

26. REISCHL, G. P., JOHN, W. and DEVOR, W. Uniform electrical charging of monodisperse aerosols. *J. Aerosol Sci.*, 1977, 8, 55–66.

27. BROWN, R. C. and WAKE, D. Quantitative measurement of the effect of electric forces in respirator filter materials. *International Symposium on Air Pollution Abatement by Air Filtration and Related Methods (Respiratory Protection)*, Dantest, Copenhagen, 1985.

28. BAUMGARTNER, H. and LOEFFLER, F. Capture of submicrometer aerosol particles in air filters by electrostatically charged fibres—An experimental and theoretical study. *Proc. 3rd Int. Conf. Electrostatic Precipitation, Abano, Venice*, 1987.

29. BAUMGARTNER, H. and LOEFFLER, F. Abscheidung submikroskopisher Partikeln mit Tiefenfiltern aus elektrisch geladen Fasern (Elektretfilter). *Staub Reinhalt. Luft.*, 1988, 48, 131–138 (in German).

30. BAUMGARTNER, H. and LOEFFLER, F. The collection performance of electret filters in the particle size range 10 nm–10 μm. *J. Aerosol Sci.*, 1986, 17, 438–445.

31. BAUMGARTNER, H. and LOEFFLER, F. A basic theoretical and experimental study of particle collection in electret fibre filters. *5th World Filtration Congress, Ostende*, 2.11–2.22, Filtration Society, London, 1986.

32. KANAOKA, C., EMI, H., OTANI, Y. and IIYAMA, T. Effect of charging state of particles on electret filtration. *Aerosol Sci. Technol.*, 1987, 7, 1–13.

33. LATHRACHE, R. and FISSAN, H. Enhancement of particle deposition in filters due to electrostatic effects. *5th World Filtration Congress Proceedings, Ostende*, 7.55–7.63, Filtration Society, London, 1986.

34. LATHRACHE, R., FISSAN, H. and NEUMANN, S. Dynamic behaviour of electrostatically charged fibrous filters. *Aerosols: Formation and Reactivity, 2nd Int. Aerosol Conf., Berlin*, 712–715, Pergamon, 1986.

35. LATHRACHE, R., FISSAN, H. and NEUMANN, S. Deposition of submicron particles on electrically charged fibres. *J. Aerosol Sci.*, 1986, 17, 446–449.

36. LATHRACHE, R. and FISSAN, H. Grundlegende Untersuchungen zum Abscheideverhalten det Elektret-Filter Teil 2: Bewertung der Filtrationseigenschaften. *Staub Reinhalt. Luft.*, 1989, 49, 365–370.

37. LATHRACHE, R. and FISSAN, H. Fractional penetration for electrostatically charged fibrous filters in the submicron particle range. *Part. Charact.*, 1986, 3, 74–80.

38. TROTTIER, R. A. and BROWN, R. C. The effect of aerosol charge and filter charge on the filtration efficiency of submicrometre aerosols. *J. Aerosol. Sci.*, 1990, 21, S689–S692.

39. EMI, H., KANAOKA, C., OTANI, Y. and IIYAMA, T. Most penetrating particle size in electret fibre filtration. In *Aerosols* (ed. B. Y. H. LIU, D. Y. H. PUI and H. J. FISSAN), Elsevier, 1984.

40. FISSAN, H. J. and NEUMANN, S. Electrostatically enhanced filtration. In *Aerosols* (ed. B. Y. H. LIU, D. Y. H. PUI and H. J. FISSAN), Elsevier, 1984.

41. FARDI, B. and LIU, B. Y. H. Performance of disposable respirators. *Part. Size Charact.*, 1991, **8**, 308–312.

42. FJELD, R. A. and OWENS, T. M. The effect of particle charge on penetration in an electret filter. *IEEE Trans. Ind. Appl.*, 1988, **24**, 725–731.

43. ANDO, K., TAKAHASHI, M., TOGASHI, R. and OKUMARA, Y. Properties of electret filter with low pressure drop and high collection efficiency. *Proc. 3rd Int. Aerosol Conf.*, Kyoto, Pergamon, 1990.

44. BROWN, R. C. Electrical effects in dust filters. *Proceedings 2nd World Filtration Congress*, 1979, Filtration Society, London, 1979, 291–301.

45. ZEBEL, G. Deposition of aerosol flowing past a cylindrical fibre in a uniform electric field. *J. Coll. Int. Sci.*, 1965, **20**, 522– 543.

46. ARIMAN, T. and TANG, L. Collection of aerosol particles by fabric filters in an electrostatic field. *Atmos. Env.*, 1976, **10**, 205–210.

47. HENRY, F. and ARIMAN, T. Cell model of aerosol collection by fibrous filters in an electrostatic field. *J. Aerosol Sci.*, 1981, **12**(2), 91–103.

48. BANKS, D. O. and KUROWSKI, G. J. Inertial efficiency of cylindrical collectors at an angle to the mean direction of flow. *Aerosol Sci. Technol.*, 1990, **12**, 312–318.

49. RAO, N. and FAGHRI, M. Computer modelling of electrical enhancement in fibrous filters. *Aerosol Sci. Technol.*, 1990, **13**(2), 127–143.

50. ZEBEL, G. Aerosol deposition on a single fibre under the influence of electrical forces 1. Theory allowing for diffusion. *Staub Reinhalt. Luft.*, 1969, **29**(2), 21–27 (in English).

51. GRADON, L. Influence of electrostatic interactions and slip effect on aerosol filtration efficiency in fibre filters. *Ind. Eng. Chem. Res.*, 1987, **36**, 306–311.

52. HOCHRAINER, D. Aerosol deposition on a single fibre under the influence of electrical forces 11. Experiments for rendering particle trajectories visible. *Staub Reinhalt. Luft.*, 1969, **29**(2), 28–32.

53. HOCHRAINER, D., HIDY, G. M. and ZEBEL, G. Creeping motion of charged particles around a cylinder in an electric field. *J. Coll. Int. Sci.*, 1969, **30**(4), 553–567.

54. KIRSCH, A. A. The influence of an external electric field on the depositiion of aerosols in fibrous filters. *Aerosol Science*, 1972, **3**(2), 25–29.

55. OAK, M. J., SAVILLE, D. A. and LAMB, G. E. R. Particle capture on fibres in strong electric fields 1. Experimental studies of filter charge fibre configuration and dendritic structure. *J. Coll. Int. Sci.*, 1985, **106**(2), 490–501.

56. OAK, M. J. and SAVILLE, D. A. Particle capture in fibres in strong electric fields 11. An experimental test of a theory for the capture of uncharged particles on charged fibres. *J. Coll. Int. Sci.*, 1985, **106**(2), 502–512.

57. IINOYA, K. and MAKINO, K. Application of electric field effects to dust collection filters. *Aerosol Science*, 1974, **5**, 357– 372.

58. NELSON, G. O., BERGMANN, W., MILLER, H. H., TAYLOR, R. D., RICHARDS, C. P. and BIERMANN, A. H. Enhancement of air filtration using electric fields. *Am. Ind. Hyg. Assoc. J.*, 1978, **39**, 472–479.

59. BERGMAN, W. and JAEGER, R. Electric air filtration movie. In *Aerosols* (ed. B. Y. H. LIU, D. Y. H. PUI and H. FISSAN), pp. 593– 595. Elsevier, 1984.

60. THURMER, H. Einige Eigenschaften grossfasserriger Filter in einem Elektrischen Hilfsfeld. *Staub Reinhalt. Luft.*, 1989, **49**, 235–240.

61. HENRY, F. S. and ARIMAN, T. Numerical modelling of electrically enhanced fibrous filtration. In *Fluid Filtration: Gas* (ed. R. R. RABER), Volume 1. American Society for Testing and Materials, Philadelphia, 1986.

62. VANOSDELL, D. W. and DONOVAN, R. P. Electrostatic enhancement of fabric filtration. In *Fluid Filtration: Gas* (ed. R. R. RABER), Volume 1. American Society for Testing and Materials, Philadelphia, 1986.

63. CUCU, D. and LIPPOLD, H. J. Electric simulated HEPA filters for particulates 0.1 μm. In *Aerosols* (ed. B. Y. H. LIU, D. Y. H. PUI and H. J. FISSAN), Elsevier, 1984.

64. CUCU, D. A new concept in the electrostatic separation of undermicronic aerosols. *Aerosols: Formation and Reactivity, 2nd International Conference, Berlin*, 1986, 723–725, 1986.

65. HELFRITCH, D. J. Performance of an electrostatically aided fabric filter. *Chem. Eng. Prog.*, 1977, **73**, 54–57.

66. YAMAMOTO, T., ENSOR, D., VINER, A. J., MOSLEY, R. B., HOVIS, C. S. and PLATTS, N. Model study for advanced electrically stimulated fabric filters. In *Aerosols: Formation and Reactivity, 2nd International Aerosol Conference, Berlin*, 1986, 719–722, 1986.

67. ZHAO, Z. M., TARDOS, G. I. and PFEFFER, R. Separation of airborne dust in electrostatically enhanced fibrous filters. *Chem. Eng. Comm.*, 1991, **108**, 307–322.

CHAPTER 7

Particle Adhesion and Particle Bounce

Introduction

Most of the theory and interpretation of experiments described so far has assumed that particles adhere to fibres on contact, but the possibility of impact without capture must now be considered. The problem of adhesion or bounce is complicated, and the amount of insight that relatively simple arguments will give is limited, but it is, nevertheless, a worthwhile exercise.

Elementary description of capture mechanism and bounce probability

As a particle strikes a fibre a complicated elastic/plastic process occurs, usually involving the dissipation of energy and possibly resulting in the particle bouncing with reduced kinetic energy. The process is quantified by the coefficient of restitution, e_r, which is the absolute value of the quotient of the velocity, V_o immediately after impact and the velocity, V_i, immediately before.

$$V_o = e_r V_i \tag{7.1}$$

In a system uncomplicated by interactions other than those occurring at and as a result of impact, bouncing will occur provided that e_r is non-zero. However, a particle close to a filter fibre is subject to forces caused by the airflow and to forces acting at a distance between the particle and the fibre. As a result, though the coefficient of restitution may be non-zero, the kinetic energy of a particle after impact may be insufficient for it to escape from the fibre. An imperfect analogy is that of a ball dropped to the ground. It will bounce but its velocity after impact will be insufficient to allow it to escape from gravity, and so it will eventually come to rest on the ground.

In equation 7.1 the approach velocity, V_i, will depend on the nature of the airflow and on the capture mechanism operating. If a particle contacts the filter fibre by interception, its velocity will be equal to that of the air carrying it. Close to the fibre that velocity is low, and so the conditions of impact will be gentle. If the particle does bounce, V_o will be low and, since the stopping distance of particles in interception conditions is short (*vid sup*), the particle will be unlikely to leave the surface. A relatively low particle–surface interaction will be sufficient to cause adhesion. As Fig. 7.1 shows, the unperturbed fluid flow at the position of the particle exerts a small force tending to keep the particle adhering to the fibre.

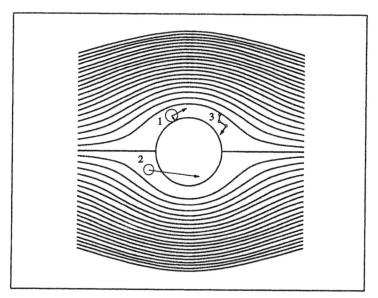

FIG. 7.1 Illustration of particle impact conditions: (1) particle captured by interception, illustrating the components of air velocity acting on the captured particle; (2) particle approaching the fibre by inertial impaction; (3) particle approaching the fibre by diffusional motion.

A particle that comes into contact with the fibre by inertial impaction will have a higher approach velocity, equal, in the limiting case, to the macroscopic air velocity. After impact it will have a higher chance of leaving the surface than the particle considered above because both its velocity and its stopping distance will be larger. Whether or not it actually will leave the surface will depend on the velocity it has immediately after impact, which will be fixed by the coefficient of restitution.

Particles that are captured by diffusional deposition will approach at the finite velocity of their own thermal motion which may take any value within a wide range. Either bounce or capture could occur, but on this scale energy dissipation has a different form from that associated with impaction, for energy is dissipated into modes rather like that of the diffusion itself. The kinetic energy of a massive particle being impacted on to the fibre is much larger than its thermal energy, and so its loss of energy is an irreversible process; capture means capture once and for all. The loss of a particle's thermal energy is reversible, and this energy may be regained from the thermal energy of the fibre. The particle will, however, be retained provided that the binding energy between the fibre and the particle is significantly larger than the thermal energy. Particles approaching the fibre by diffusion and impaction are illustrated in Fig. 7.1.

Particles that are attracted to the fibre by electrostatic forces will tend to strike the fibre at a velocity equal to the drift velocity caused by the electric field, along with any inertial or diffusional component. If the coefficient of restitution is a constant, the higher impact velocity will result in more energy loss, and the remaining energy may be insufficient for escape from the electric forces, which will act to hold the particle close to the fibre. The net effect will be that these forces will make adhesion more likely.

Relative importance of bounce and re-entrainment

If a particle is observed to leave a fibre after it has made contact a question may be asked as to whether the cause of its release is bounce or capture followed by subsequent re-entrainment. At the moment of impact of a particle the force tending to dislodge it is not likely to be very different from that acting on a particle already captured, but there is in addition the energy of restitution available to aid its dislodgement. Detachment is, therefore, more likely to occur at the moment of impact than subsequently, and it can be reasonably concluded that particles that do not bounce are unlikely to be re-entrained by the airflow from which they were captured.

Re-suspension of particles by the airstream or by particle–particle impact

Although simple re-entrainment of particles is not likely, the situation may be different if the air velocity or other conditions are altered. In the Stokes flow regime, an increase in the macroscopic air velocity means a proportional increase in the microscopic velocity close to the fibre surface. At still higher velocities, Stokes flow gives way to inertial flow, and the velocity close to the surface is increased relatively more; the drag on the captured particles increases and re-entrainment becomes more likely. Conditions may be further altered if the direction of airflow is reversed, because this will change the direction of the small normal component of the drag force, making it act on the particle in a direction away from the fibre.

A further effect, occurring only when fibres are heavily loaded, is the bombardment of captured particles or clusters by approaching airborne particles. Particle–particle impact may be different in nature from particle–fibre impact, and the transfer of momentum could result in the detachment of particles in clusters, making the filter act as an agglomerator. Finally, the detachment of particles may be aided by physical vibration, though different modes of vibration may vary in their effectiveness in removing captured particles.

Adhesion forces between particles and fibres

It is useful to consider first the static forces between captured particles and fibres, for these are sufficient to describe the behaviour of particles once they have been captured. A study of bounce involves not only these but also the dynamics of impact, and this will be made later. The three principal forces attaching particles to fibres are van der Waals forces, surface tension and electrostatic forces, though adhesion can be influenced by any sort of interaction (1) between two bodies, and it is complicated by the effects of contaminants.

Van der Waals forces

Van der Waals forces between atoms arise because an atom has a fluctuating electric dipole moment, the electric field of which will induce and then attract a dipole in a neighbouring atom. The process is rather like the induced dipole force on aerosol particles described in Chapter 6, but the field, which in this case arises from a point dipole rather than a line dipole, is shorter in range, varying as r^{-7}. The energy of interaction by this process is,

$$\Phi_{\text{vw}} = -\frac{\beta_1}{r^6} \qquad (7.2)$$

where β_1 is related to the atomic polarisability of the atom and its ionisation potential (2).

In macroscopic bodies, the interaction between any two atoms will be complicated by the dielectric effect of atoms in between. The result of a detailed treatment gives, for the attractive force between a sphere of radius, R, and a plane at a distance, z_0, from it (1),

$$F = \frac{h v R}{8 \pi z_0^2}.$$ (7.3)

The energy hv, the Lifshitz van der Waals constant (1) corresponding to the ionisation potential, is a fundamental property of the material considered, and varies between about 0.6 eV for insulating polymers of the sort used in fibrous filters and 2–10 eV for semi-conductors or metals. If the two materials interacting are different, hv is replaced by the geometric mean of the values appropriate to each. In the case of an irregularly shaped particle R is replaced by R_a, the radius of the surface asperity that is closest to the plane.

The value of the force depends critically on the separation distance z_0, which is often taken to be 0.4 nm, a distance slightly larger than the lattice spacing of van der Waals crystals and, therefore, the order of the equilibrium atomic separation obtained with van der Waals forces.

Van der Waals forces are attractive at any distance, but the stable conditions implied by specifying z_0 can occur only if there is a repulsive force as well; and a good empirical potential that includes this is the Lennard-Jones potential, illustrated in Fig. 7.2 and given by,

$$\Phi_{LJ} = \frac{\beta_2}{r^{12}} - \frac{\beta_1}{r^6}.$$ (7.4)

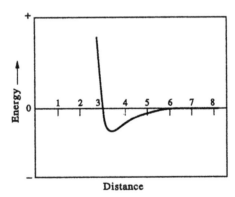

FIG. 7.2   Lennard–Jones potential (equation 7.4).

Strictly speaking the forces between bodies acting at a distance are not instantaneous but are transmitted at the velocity of light, which is significant for van der Waals forces because the dipole fluctuations are rapid. The effect of this would be to reduce the range of the forces, but the effect can be neglected provided that the position of closest approach of the bodies is less than a few tens of nm.

Normally van der Waals forces between macroscopic bodies are expressed in terms of the Hamaker constant, A_1 (3).

$$F = \frac{A_1 R_a}{6 z_0^2} \tag{7.5}$$

A list is given in Table 7.1 (3), of Hamaker constants for metallic and non-metallic materials.

TABLE 7.1
Hamaker constants (3)

Material	Hamaker constant (10^{-20} J)
Calcium oxide	12.4
Carbon (diamond)	28.4
Carbon (graphite)	46.9
Copper	28.3
Gold	45.4
Lead	21.4
Magnesium oxide	10.6
Mica	10.0
Nylon	30.0
PSL	6.37
PTFE	60.0
PVA	8.84
Silica	8.53
Silver	39.80
Steel	21.20
Zinc	21.80

Surface tension (capillary) forces

Figure 7.3 shows a sphere attached to a plane by means of a liquid bridge. If the angle of contact is zero, the force between them is (4),

$$F = 4 \pi \tau_T R_a \tag{7.6}$$

where τ_T is the surface tension of the liquid, and R_a is the particle or roughness radius. A calculation based on simple geometry shows that the force is independent of the amount of liquid present provided that a complete bridge is formed, because as the area of contact diminishes the curvature increases, and so does the negative internal pressure of the

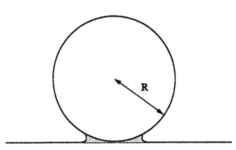

FIG. 7.3   Sphere attached to a plane by capillary forces. The surface tension force is independent of the particle radius, R.

liquid bridge. This constancy depends, of course, on the particle being spherical, and it will not hold for particles of all shapes.

Electric forces

Particles may become charged by triboelectric exchange with the fibre, or they may be charged before they touch the fibre. In the latter situation, where the particle is a non-conducting sphere with charge, q, and radius, R, the force of interaction can be calculated if the approximation is made that the fibre is replaced by a plane. This approximation is reasonable provided that the diameter of the particle is much less than that of the fibre, having been used successfully in Chapter 6, when image forces were discussed. The approximation of replacing the fibre by a plane is frequently applied in experimental measurements of adhesion, for many of these are carried out using planar substrates rather than fibres.

The expression for the force is different from that used in Chapter 6, since the charge on the particle was then assumed to be a point charge at its centre. This approximation is suitable for action at a distance, but when adhesion is considered, and particle–fibre contact has taken place, it is no longer valid. The correct expression for the force of adhesion due to electric charge is (4),

$$F = \frac{q^2}{16\pi\varepsilon_0 R\,\delta} \frac{\ln\left(1 + \dfrac{\delta}{z_0}\right)}{\left[\gamma + \frac{1}{2}\ln\left(\dfrac{2R}{z_0}\right)\right]\left[\gamma + \frac{1}{2}\ln\left(\dfrac{2R}{z_0 + \delta}\right)\right]} \qquad (7.7)$$

where δ is the depth at which the charge density falls to e^{-1} of that at the surface, γ is Euler's constant, and z_0 is, as above, the distance between the sphere and the plane. Equation 7.7 reverts to a simpler form if both the particle and the plane are conductors, because $\delta = 0$ (1).

Relative magnitude of adhesion forces

Accurate calculation and prediction requires a knowledge of the geometry and other properties of the system, that is not usually available; but a comparison of the magnitude of the different forces acting on a specimen system comprising a 10 μm diameter quartz particle and a polymer fibre has been made (4). The distance of closest approach is assumed to be 0.4 nm, and the particle has a three-point contact, with surface asperities 0.1 μm in diameter. The Lifshitz van der Waals constant is assumed to be 2 eV, a reasonable estimate of the geometric mean of those for quartz and a typical polymer, and so the attractive force due to van der Waals interactions will be approximately 2×10^{-8} N. For the same particle holding 1000 elementary charges, and with $\delta = 0.5$ nm, the electrostatic force will be 5×10^{-10} N, and for capillary forces acting at the three asperities described above, the force will be 3×10^{-7} N.

The relationship in equation 7.6 and the importance of surface tension forces has been demonstrated with large particles at 100% RH, but at lower humidity the force is considerably diminished (5, 6). Experiments carried out with glass spheres adhering to a glass slide, showed the force to vary little as the humidity was reduced to about 80%, and then to drop very rapidly, probably as a result of imperfect wetting. The force is also significantly reduced in practice by the presence of hydrophobic contaminants.

The action of electric forces in adhesion has been demonstrated, again on rather large particles (of fly ash) attracted to acrylic fibres (7) in experiments during which the fibres were made to vibrate, causing the particles to be shaken off. The detached particles were observed to describe orbits around the fibres, indicating the existence of a long-range force, which can only be identified with electric force. These findings confirmed that electric forces influenced adhesion; but the forces required to cause detachment of particles were much greater than the forces acting to keep the particles in orbit, indicating the presence of a much larger short-range force, which was identified with the van der Waals force.

It is possible that electric forces, since their range is considerably longer than that of the other adhesion forces, are important in securing the adhesion of pre-charged particles, allowing the van der Waals force to maintain the adhesion. The electric forces may not persist if both the particle and the fibre are conducting. Likewise capillary forces, though sometimes large, are dependent on conditions. Van der Waals forces are ever-present and these are concluded to be the most important adhesion forces.

In the work that follows, the measured values of adhesion energy, or the values implied by the behaviour of the system under study, are frequently larger than the simple theory would predict. One explanation is the occurrence of inelastic deformations at the points of contact, increasing the strength of attraction by a time-dependent factor which varies with the extent to which the softer of the two components can be deformed (8).

Elastic deformation of both particle and fibre will also occur, but the proper treatment of this and of the effects that it produces are matters of controversy (3, 9, 10, 11). That adhesion energy is larger than simple theory predicts is, however, not disputed.

Range of values of adhesion forces

Many measurements of adhesion have been carried out on planar rather than fibrous substrates, but the results can be generalised. Probably the most favoured method of measuring the adhesion of particles to surfaces is the application of extremely high forces (up to 10^5 g or more (12)) by centrifuging the particles and their collector, and counting the particles remaining after a period of exposure to the detaching force. An alternative method is detachment by a high velocity air stream, but in this situation it is not so easy to calculate the force acting. Moreover it has been observed (13) that whereas protracted exposure of deposits to centrifuging does not result in increased particle detachment, prolonged use of high velocity air does, possibly as a result of turbulence. A further measurement technique is the use of ultrasound, by which forces as high as those given by centrifugation can be applied, though the force is intermittent whereas that of centrifugation is constant. Vibration of a loaded substrate often needs to be carried out over a period of several seconds before significant particle detachment occurs.

Centrifugation experiments carried out on particles of starch a few micrometres in diameter attached to a starch-covered substrate, and on particles of iron attached to iron (12), showed a wide range of effective adhesion energies, a factor of ten or more existing between the force that caused detachment of 10% of the particles and that which caused 90% to be detached. The reason put forward for this was that although all particles may have the same average surface structure, on a microscopic level the surfaces actually in contact with the collector may be geometrically very varied. Forces of up to 10^{-5} N were needed to detach the starch particles, but the material is a well-known adhesive. The iron could be removed by forces of about 10^{-7} N.

The same method was used (4, 14) to study particles of silica captured by impaction on a polyamide fibre, giving adhesive forces very similar to those of the iron. The distribution of adhesion energies was well-described by a lognormal function, as shown in Fig. 7.4, and the geometric standard deviation was the same irrespective of the particle size, and similar to that for the iron particles above.

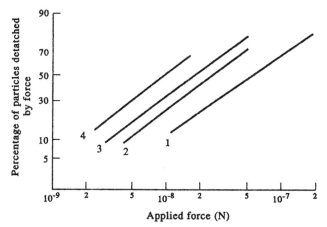

FIG. 7.4 Distribution of adhesion energies of quartz particles deposited at a filtration velocity of 0.42 ms^{-1}, on polyamide fibres; (1) 15.1 μm particles; (2) 10.3 μm particles; (3) 8.3 μm particles; (4) 5.1 μm particles (4). (With permission from Springer-Verlag, Heidelberg.)

In experiments carried out on silica particles attached to polyamide fibres (14) in which the particles were detached by air at high velocities, a factor of ten or more existed between the air velocity that would remove 10% of the particles and that which would remove 90%, the same observation having been made in earlier work (8) involving the blow-off of particles of silica deposited on wire meshes.

A wide range of adhesion energies is generally observed amongst nominally identical particles on a single substrate. A useful parameter quantifying the binding force is the force required to detach 50% of the particles of any specific type from a specified substrate (8).

In the case of particles with a very low adhesion energy, such as small particles deposited by diffusional deposition, the application of a dislodging force for a finite time should remove some but not all (15). The number of identically bound particles adhering to a surface decreases exponentially with time, since this function describes their probability of obtaining sufficient thermal energy to cause them to become detached. The time constant could be very large, depending on the binding energy between the particles and the fibre, but the application of an additional force tending to cause detachment would reduce this. There appear to be no experimental measurements of this effect.

Effect of particle and fibre properties

A wide range of properties of both the particles and the fibre might contribute to the strength of the adhesion. Experimenters do not agree entirely, and a serious practical problem in many experiments is to isolate a single effect. Furthermore, simple and useful dimensionless parameters describing this process have not been defined.

Adhesion forces are stronger when acting on large particles, as shown in Fig. 7.4 (6, 12), but the drag exerted on them by an airflow or the force exerted on them by centrifugation is also greater; and large particles or agglomerates can be detached by a lower airflow (16) or angular velocity than small. It is also easier to detach particles from thick fibres than from thin ones (6).

Hardness is important in adhesion but, as stated above, it is the hardness of the softer adhesion partner, usually the fibre, that is critical (17). The degree of adhesion to soft fibres does not appear to depend on the hardness of the material constituting the particle (4). In experiments on three types of fibres with the same aerosol, adhesion was found to be greatest for polyamide, followed by glass, followed by polyester, the observed behaviour of the fibres possibly being affected by absorbed water influencing their hardness.

The cross-sectional shape of the fibre has a small but measurable effect, and it has been observed (16) that deposits on trilobal fibres needed air velocities about 20% higher for their removal than did deposits on fibres with circular cross sections.

Treating either the fibre or the particles to improve the surface smoothness and, therefore, the degree of contact between the surface and the particle, increases the strength of the adhesion. In the case of adhesion of iron particles to iron substrates discussed above (11), chemical reduction of either component to improve surface smoothness increased the adhesion efficiency, and the greatest adhesion was achieved when both were chemically reduced. In other work glass spheres, with their smooth surfaces, were found to be attached more tenaciously to polyamide fibres than were silica or limestone (16).

The effect of particle shape has been investigated by using ultrasonic vibration to detach particles of similar chemical composition but a variety of shapes (18). The observed adhesion forces correlated well with the expected area of contact, being the largest for flakes, second for fibres, third for spheres and smallest for irregular particles with relatively few points of contact.

Influence of filtration conditions

The adhesion of particles to fibres is greater for particles that have been captured at a higher filtration velocity. Although a high filtration velocity does, in conditions of inertial impaction, make bounce more likely, the particles that are captured are particularly tightly bound, and often very difficult to remove.

The effects are illustrated in experimental work using the centrifuge method (14). With the force of adhesion defined as the force required to remove 50% of the particles, 5 μm particles of quartz were observed to be approximately twice as tightly bound to a polyamide fibre by which they were captured at a filtration velocity of 0.63 ms^{-1} than were similar particles that were just allowed to settle onto the fibre. The adhesion force for particles captured at intermediate velocities increased monotonically with the velocity. The factorial difference increased steadily with particle size, up to a factor of about 5 between the adhesion forces for 15 μm particles captured at 0.63 ms^{-1} and those captured by settling. A similar pattern was observed with limestone particles.

It has been stated above that the air velocity required to re-entrain particles is several times the filtration velocity. Experimental measurements of particle detachment by high air speeds are consistent with the centrifuge measurements above, in that the air velocity needed to remove 50% of the particles is 5–10 times the filtration velocity, whatever the latter's value might be.

Increasing relative humidity tends to improve particle adhesion. In early experiments with large particles attached to large dry fibres, the air velocity required to remove the particles at a relative humidity of 40% was about three times greater than that needed at a RH of 22% (6).

In other experiments (4, 13) the effect of relative humidity on the adhesion of quartz particles to fibres of polyamide, polyester and glass was investigated. Adhesion was very similar on all three fibres at relative humidities below about 40%, but at 80% RH adhesion to glass had increased threefold and that on the others by about 50%, with a higher value for polyester. Glass has the lowest moisture regain of the three fibres and polyamide the highest. The magnitude of the effect was significantly lower than the theoretical predictions based on equation 7.6, which describes capillary forces. This is consistent with the observation that capillary forces are significant at high relative humidity only (5). An alternative explanation of what was observed is that the fibres were slightly softer at the higher humidity, which would influence the dynamics of impact. The form of variation indicated against significant electrostatic attraction, because this would have been reduced at high relative humidity.

Adhesion to a fibre with a fixed coating of the particulate material, simulating a partly loaded fibre, tends to be weaker than adhesion to the fibre itself. Quartz particles attached to other quartz particles were observed (4) to have only about half the adhesion force of similar particles attached to a polyamide fibre, which can be explained in terms of a hardness effect, with the hypothesis that it is the hardness of the softer component that is critical (1).

Dynamics of impact

The theory of impact is less straightforward than that of adhesion because it cannot be described solely in terms of adhesion forces, which are static entities; it depends critically on the dissipation of energy.

Dissipation can come about by intermediate means such as the propagation of elastic waves, with the energy ultimately appearing as heat, which is kinetic energy distributed irrecoverably on a molecular level (an increase in entropy). Alternatively the energy can be spent in doing work to effect plastic deformation of the fibres, also resulting in irreversible change. In the situation where an attractive force acts between the particle and the fibre the viscous drag of the air aids capture by dissipation of energy.

In the impact of smooth spherical particles at normal incidence on a smooth surface, the velocities of the particles before and after collision are related by means of equation 7.1. At the point of rebound the kinetic energy of the particle, which is the sum of the energy imparted by the flow field and by attractive forces acting prior to contact, is reduced, being scaled by a factor of e_r^2. As it leaves the surface the particle must now overcome the energy of adhesion Φ_r and so its final energy will be given by subtracting this quantity. A complication is that the energy of adhesion for the rebounding particle may be different from that for the approaching particle because the physical nature of the system may have altered. For example the particle or fibre may have been deformed plastically, or triboelectric exchange of charge may have taken place. In the simplest situation, where the two adhesion energies are the same, the critical value of the approach velocity, that at which there will be just sufficient energy for the particle to escape, will be given by (19),

$$V_i = \left[\frac{2\Phi_r(1 - e_r^2)}{me_r^2} \right]^{1/2}. \qquad (7.8)$$

In the other extreme, where the adhesion energy for rebound is much greater than that for approach, the critical velocity will be,

$$V_i = \left[\frac{2\Phi_r}{me_r^2} \right]^{1/2}. \qquad (7.9)$$

In the conditions described by equation 7.9 particles may be captured even if $e_r = 1$, but this does not apply in conditions described by equation 7.8. Both equations are strictly correct, but their solution requires that the values of all of the parameters contained, in particular the coefficient of restitution, be known.

The relationship between velocities before and after impact, implicit in equation 7.8 has been demonstrated experimentally on planar surfaces (20, 21), and is illustrated in Fig. 7.5. From the results it is possible to calculate the adhesion energy, and this is, as mentioned above, found to be considerably greater than the theoretical prediction based on equation 7.3 for spherical particles, but if flattening of the particles on impact takes place the area of contact, and therefore the force of interaction, can be considerably increased.

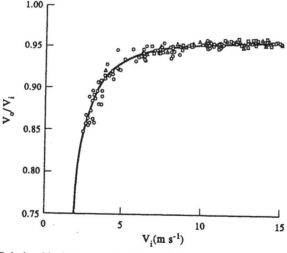

FIG. 7.5 Relationship between velocities of 1.27 μm polystyrene latex particles before and after impact on a quartz target. The solid line is fitted to the data with e_o and ϕ_r as variables (21). (With permission from Academic Press.)

A significant component of the coefficient of restitution, e_o, is independent of the geometry of impacting bodies, and due principally to internal friction, but there is in addition a component caused by flexural work on the fibre (22, 23). A function giving a good fit to the calculations of the coefficient of restitution including this effect is,

$$e_r = e_o + \exp\{-1.7e'(N_R)\} - 1 \qquad (7.10)$$

where e', calculated on the basis of elasticity theory (23, 19), varies approximately quadratically with the interception parameter as defined in equation 4.5.

Face velocity and impact velocity

Equations 7.8 and 7.9 show that the adhesion probability depends on the kinetic energy of impact of a particle, which is related to but not identical with the kinetic energy that the particle would have if it were travelling at the face velocity of the filtered air. A simple correlation between adhesion probability and kinetic energy before impact has been carried out (24), including the results of several workers (17, 25), illustrated in Fig. 7.6, and supplemented by later work (26). Correlation is clear, but there is also considerable spread in the results, caused partly by the fact that face velocity is not the same as impact velocity (*vid inf*) and partly by the fact that the simple theory leaves a lot of significant parameters unaccounted for.

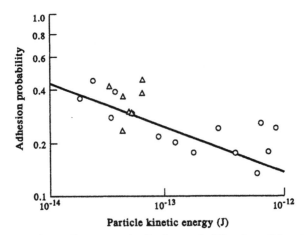

FIG. 7.6 Experimentally measured adhesion probability for solid particles as a function of kinetic energy. $\triangle$, reference 26; $\bigcirc$, reference 24; $-$, references 17 and 25.
(© 1987, Elsevier Science Publishing Co. Inc.)

The relationship between critical velocity and bounce probability following from this relationship and equations 7.8 and 7.9 gives calculated critical velocities that appear to be rather low. However, the flow field will slow down the particle in the region of stagnation surrounding the fibre, and so the face velocity at the onset of bounce will be rather higher than the critical impact velocity.

The distinction between air velocity and impact velocity was made in work (27) interpreting the results of filtration of air containing felspar dust by metal wire filters. In order to calculate the probability of bounce the impact velocity was calculated by the numerical method outlined in Chapter 4, for the Kuwabara (28) flow field. Since the impact velocity varies critically with Stokes number over a small range of values, the effect of *St* on bounce probability is considerable. The calculated Stokes number at which 50% of the particles bounce was found to be within a factor of about two of that observed experimentally.

A factor mentioned in Chapter 4 though not treated in detail is the hydrodynamic resistance of the cushion of air between an impacting particle and the collecting fibre. This may influence the impact conditions (29) by making the contact more gentle, but the effect will be reduced by any aerodynamic slip (30).

Factors influencing bounce probability

A brief review of the static and dynamic factors influencing the extent of bounce (31) named contact area, which is influenced by natural properties such as hardness, particle shape, surface roughness, and both the angle and velocity of impact. Relative humidity is a significant factor because it may alter the hardness of the particles or the fibres, and affect dynamic behaviour as it does static (4). The action of moisture through capillary effects and the screening of any electric field have both been mentioned above, and a further process is the shielding of van der Waals forces (4) which would reduce adhesion; though the experimental results (32) all show an increase in impact efficiency with increasing relative humidity.

Effect of oil on the fibre surface

A well-known method of increasing the collection efficiency of filters is to coat the fibres with oil. Experiments (33) performed on extremely thick (220 μm) metal fibres, illustrated that improvement was obtained with a variety of oils, but olive oil and sunflower oil proved the most effective of those used and castor oil rather less so. The various oils varied little in surface tension but considerably in viscosity, and the less viscous oils were more effective, as also observed in experiments with oiled planar surfaces (34). The use of vaseline, with its extremely high viscosity and non-Newtonian behaviour did not give good performance, though fibres coated with vaseline were more effective than fibres used dry. In previous work (35) olive oil was found to be superior to glycerol.

Over short periods of loading the thickness of an oil coating has little effect. In experiments on meshes, a coating of 75.7 g/sq metre had no more effect than one of 20.4 g/sq metre on the initial adhesion efficiency. As the meshes became loaded the efficiency of both fell off, but the point of onset of the reduction in efficiency was at a higher load for the more highly oiled mesh, and at all loadings higher than this the more heavily oiled mesh was more efficient. It seems likely that at high loads the oil on the heavily oiled fibre is able to seep through the particle deposit, making the surface sufficiently oily to improve the filtration efficiency.

Clean air at very high velocity passed over heavily oiled and dust-loaded fibres carried the deposit to the rear of the fibre where its detachment was aided. The reason for this was that the oil, which caused the particles to adhere, also reduced friction between them and the fibre. Oiling had the benefit, however, that whereas bombardment of captured particles by airborne particles larger than about 5 μm sometimes caused detachment from dry fibres, it did not from oiled ones.

Viscous oil, though less effective in reducing bounce, was more effective in preventing detachment of the particles once captured, one of a number of instances where a physical parameter reduces both adhesion and bounce probability.

Effect of particle size and shape

Small particles are generally observed to be less likely to bounce than are large particles, whatever the system under study (36, 37, 38, 39). Particle shape also affects bounce, but in a different way from its effect on adhesion. Irregularly shaped particles of quartz (36, 37) are less likely to bounce than smooth spheres of glass (in contrast with static adhesion experiments), the interpretation given of this being that the surface asperities are more likely to undergo plastic deformation. Similarly (39) glass spheres in impactors were observed to bounce more than ammonium fluorescein particles, and both of these more than naturally occurring particles of pollen.

Effect of collector surface

Coarse fibres are found to result in less bounce than fine fibres (36) if the velocity of impact is kept constant. In experiments using planar collectors of different materials (39) it was found that stainless steel caused less bounce than aluminium, and both less than cellulose acetate.

Although static measurements have shown that the adhesion of particles to a layer of similar particles is weaker than the adhesion to a clean fibre, the capture efficiency of a fibre with a deposited layer is actually greater (4), behaviour also observed in impactors (39). The reason suggested for this is that the particle layer allows more dissipation of energy on impact, even though the adhesion forces are weaker. Other results (35) show particle adhesion increasing with dust loading if the initial efficiency is high, but decreasing if it is initially low. The capture of an aerosol can be augmented by the inclusion of material like soot, which can deposit on the fibres and influence their surface properties by providing a "landing pad" (31).

Critical impact velocity and direct observation of impact

An expression has been derived for the critical impact velocity, taking into account both the dissipation of energy caused by plastic deformation at impact and the increased van der Waals interaction as a result of the increased area of contact between the fibre and the particle, due to this deformation (36). The expression includes the coefficient of restitution for purely plastic deformation, e_{pl}, the microscopic yield pressure, p_{pl}, the particle density, ρ, and the van der Waals constants.

$$V_c = \frac{1}{d_p} \frac{(1 - e_{pl}^2)^{1/2}}{e_{pl}^2} \frac{A_1}{\pi z_0^2 (6 p_{pl} \rho)^{1/2}} \tag{7.11}$$

The theory treats conditions at impact. Trajectory calculations are required for a description of the particle's approach for the fibre. These agreed with the results of previous work (27) in predicting that the impact velocity is only a fraction of the face velocity, and also that it increases monotonically with angular distance from the forward stagnation point.

Experimental measurements of adhesion were obtained by high speed microphotography (37) which not only illustrated the process in action but also enabled the coordinates of the particles to be precisely defined over a series of incremental times, so that velocity and acceleration could be measured directly.

Experimental results for the adhesion probability are shown in Fig. 7.7. Theory predicts a steady drop in adhesion efficiency but the experimental results for adhesion flatten out at higher velocities.

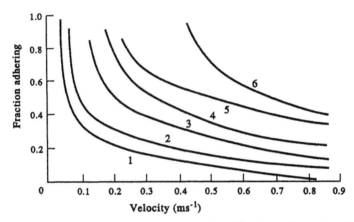

FIG. 7.7 Experimentally measured adhesion probabilities of particles to glass fibres as a function of velocity: (1) 10 μm glass spheres, 20 μm polyamide fibre; (2) 5 μm glass spheres, 20 μm polyamide fibre; (3) 10 μm quartz particles, 20 μm polyamide fibre; (4) 5 μm quartz particles, 20 μm glass fibre; (5) 5 μm quartz particles, 20 μm polyamide fibre; (6) 10 μm quartz particles, 50 μm polyamide fibre (36). (With permission from Springer-Verlag, Heidelberg.)

In complementary work (39) also involving microphotography, the effect of angle of incidence on bounce probability was investigated, using angles of 90° (normal incidence) and 45°. The bounce probability was expressed as a function of the normal component of the velocity, and it was found that normal incidence resulted in less bouncing of particles than did oblique incidence.

A further observation, from the direct visual method, was the relatively common occurrence of the dislodgement of a single captured particle as a result of the impact of a single similar approaching particle.

Adhesion and bounce of very small particles

The possibility of very small particles not agglomerating on collision because of their mean thermal velocity due to Brownian motion exceeding their capture velocity is discussed in work previously cited for another reason (19). That filtration efficiency should be limited in the same way is implicit in the work.

The smallest possible "particles" are single molecules, and whilst these, because of their high diffusivity, make frequent collisions with a surface, they do not adhere. Small molecular clusters will behave in the same way, whereas submicrometre particles deposited by diffusional deposition will have a very high adhesion probability. There should, therefore, be a range of particle sizes over which the capture efficiency drops significantly.

The mean kinetic energy of particles can be calculated from the theory of equipartition of energy:

$$\tfrac{1}{2}mV^2 = \tfrac{1}{2}k_{\mathrm{B}}T \qquad (7.12)$$

which translates readily to a mean approach velocity of,

$$V = \left(\frac{k_{\mathrm{B}}T}{m}\right)^{1/2} = \left(\frac{6k_{\mathrm{B}}T}{\pi\rho}\right)^{1/2} d_{\mathrm{p}}^{-3/2}. \qquad (7.13)$$

The velocity of impact increases as the particles become smaller and eventually the probability of bounce will dominate. There is an interesting parallel between this and the case of particles of high Stokes number. As *St* increases, the velocity of impact increases, and whilst this results in increased frequency of impact, it ultimately acts against adhesion.

A detailed theory of the bouncing of very small particles (40) involves the calculation of the critical approach velocity as a function of particle size. The functional relationship depends on the theory used, but is close to inverse proportionality; whereas the mean thermal velocity varies as the particle diameter is raised to the power $\tfrac{3}{2}$, as indicated in equation 7.13. Graphs of the two functions will cross at a point that depends on the mechanical properties of the materials considered. Since the actual thermal velocities will be distributed about the mean value, the result will be a steady decrease in collection efficiency as the particle size is reduced; and the typical particle diameter range over which significant reduction in efficiency occurs is calculated to be 10–100 nm.

The filtration efficiency can be improved by a reduction of the temperature of the system, because the reduced bounce more than compensates for the reduction in particle transport caused by the lower diffusivity at lower temperatures, indicated by equation 4.33.

Transfer of electric charge on contact

The transfer of charge between a particle and the fibre, in the form of contact or frictional electrification (41) is, like plastic deformation, a process that can result in a greater adhesion energy for particles leaving a surface than for those approaching it. The transfer depends on the nature of the surfaces and any contaminants, and on the area and time (22) of contact, both of which will increase with the velocity of impact.

This phenomenon does not appear to have been investigated in filters, but experiments have been carried out using methylene blue particles impacting on a metal surface at velocities of 100 metres per second or more (42). Generalisation to a filter cannot be accurately made because of the degree of extrapolation required, but the results suggest that the effect in most filtration systems is likely to be small; though observations of polymer–polymer contact (43) show that the charge exchanged tends to be higher in this situation than in polymer–metal contact.

Capture of liquid droplets

The largest intrinsic force acting on liquid droplets is surface tension. If the particles are very small this force is so large that it imparts a rigidity to the droplets, making them behave very much like solid particles. Larger particles behave more like a bulk liquid, and these have special properties when it comes to filtration. They are able to absorb energy by deformation and the associated internal viscous drag. In addition they may wet the fibre, enabling surface tension to aid adhesion. They are also much more likely than solid particles to break up on impact with a fibre, particularly when incidence is not normal and impact gives rise to a shearing force (44).

Fibre shedding

A final topic, more appropriate to this chapter than to any other, is the problem of fibre-shedding, whereby a filter loses some of its own fibres to the air passing through. The process resembles the re-entrainment of captured particles. It is very significant from the point of view of the high air quality required in clean rooms, and from the point of view of the health hazard of respirable fibres shed from respirators.

If a filter has a number of weakly bound fibres, these are most likely to be shed when the filter is first used (45). An initial surge of fibres is produced, which eventually decreases; and a situation is reached where air at a constant velocity is unlikely to cause further fibre detachment. If the air velocity is increased or pulsed, further detachment may take place (46). From a practical point of view, clean room filters may be purged with clean air at a relatively high velocity before they are used (47).

Fibres can also be shed if filters are subjected to vibration or shock (48), and physical manipulation of unprotected filter material can break fibres and make shedding likely (49).

References

1. KRUPP, H. Particle adhesion. Theory and experiment. *Adv. Colloid Interface Sci.*, 1967, **1**, 111–239.
2. RENNE, M. J. and NIJBOER, B. R. D. Microscopic derivation of macroscopic van der Waals forces. *Chem. Phys. Lett.*, 1967, **1**, 317– 320.
3. TSAI, C. J., PUI, D. Y. H. and LIU, B. Y. H. Elastic flattening and particle adhesion. *Aerosol Sci. Technol.*, 1991, **15**(4), 239–255.
4. LOEFFLER, F. The adhesion of dust particles to fibrous and particulate surfaces. *Staub Reinhalt. Luft.*, 1968, **28**(11), 29–37.
5. MCFARLANE, J. S. and TABOR, D. Adhesion of solids and the effect of surface films. *Proc. Roy. Soc.*, 1950, **A202**, 224–243.
6. LARSEN, R. I. The adhesion and removal of particles attached to air filter surfaces. *Ind. Hyg. Assoc. J.*, 1958, 265–270.
7. RAFAILIDIS, S. V. and REIZES, J. A. Determination of the adhesivity on acrylics of fresh coal fly ash particles. *Part. Syst. Charact.*, 1985, **5**, 130–138.
8. CORN, M. and SILVERMAN, L. Removal of solid particles from a solid surface by a turbulent air stream. *Am. Ind. Hyg. Assoc. J.*, 1961, **22**, 337–347.
9. DAHNEKE, B. The influence of flattening on the adhesion of particles. *J. Coll. Int. Sci.*, 1972, **40**(1), 1–13.
10. DERYAGUIN, B. V., MULLER, V. M. and TOPOROV, Y. P. Effect of contact deformation on the adhesion of particles. *J. Coll. Int. Sci.*, 1975, **53**(2), 314–326.
11. DERYAGUIN, B. V., MULLER, V. M., MIKHOVICH, N. S. and TOPOROV, Y. P. Influence of contact electrification on the collision of elastic particles with a rigid surface. *J. Coll. Int. Sci.*, 1987, **118**(2), 553–563.
12. BOEHME, G., KRUPP, H., RABENHORST, H. and SANDSTEDE, G. Adhesion measurements involving small particles. *Trans. Inst. Chem. Eng.*, 1962, **40**, 252–259.
13. CORN, M. and STEIN, F. Reentrainment of particles from a plane surface. *Am. Ind. Hyg. Assoc. J.*, 1965, **26**, 325–366.
14. LOEFFLER, F. Investigating adhesive forces between solid particles and fiber surfaces. *Staub Reinhalt. Luft.*, 1966, **26**(7), 10–17 (in English).
15. DAHNEKE, B. Note. Resuspension of particles. *J. Coll. Int. Sci.*, 1975, **50**(1), 194–196.
16. LOEFFLER, F. Blow-off of particles collected on filter fibres. *Filtration in Process Design and Development*, Sept. 28–30, 33–37, Filtration Society, London, 1971.
17. LOEFFLER, F. Adhesion probability in fibre filters. *Clean Air*, 1974, **8**(4), 75–78.
18. MULLINS, M. E., MICHAELS, L. P., MENON, V., LOCKE, B. and RANADE, M. B. Effect of geometry on particle adhesion. *Aerosol Sci. Technol.*, 1992, **17**, 105–118.
19. DAHNEKE, B. The capture of particles by surfaces. *J. Coll. Int. Sci.*, 1971, **37**(2), 342–353.
20. DAHNEKE, B. Measurements of bouncing of small latex spheres. *J. Coll. Int. Sci.*, 1973, **45**(3), 584–590.
21. DAHNEKE, B. Further measurements of the bouncing of small latex spheres. *J. Coll. Int. Sci.*, 1975, **51**(1), 58–65.
22. HUNTER, S. C. Energy absorbed by elastic waves during impact. *J. Mech. Phys. Solids*, 1957, **5**, 162–171.
23. ZENER, C. The intrinsic elasticity of large plates. *Phys. Rev.*, 1941, **59**, 669–673.
24. ELLENBECKER, M. J., LEITH, D. and PRICE, J. M. Impaction and particle bounce at high Stokes numbers. *J. Air Pollut. Control Assoc.*, 1980, **30**(11), 1224–1227.
25. ESMEN, N. A., ZIEGLER, P. and WHITFIELD, R. The adhesion of particles upon impaction. *J. Aerosol Sci.*, 1978, **9**, 547–556.
26. DUNN, P. F. and RENKEN, K. J. Impaction of solid aerosol particles on fine wires. *Aerosol Sci. Technol.*, 1987, **7**, 97–107.
27. STENHOUSE, J. I. T. and FRESHWATER, D. C. Particle adhesion in fibrous air filters. *Trans. Inst. Chem. Eng.*, 1976, **54**, 95–99.
28. KUWABARA, S. The forces experienced by randomly distributed parallel cylinders or spheres in viscous flow at small Reynolds numbers. *J. Phys. Soc., Japan*, 1959, **14**, 527–532.
29. GOREN, S. L. and O'NEILL, M. E. On the hydrodynamic resistance of a particle of a dilute suspension when in the neighbourhood of a large obstacle. *Chem. Eng. Sci.*, 1971, **26**, 325–328.

30. BARNOCKY, G. and DAVIS, R. H. The effect of Maxwell slip on the aerodynamic collision and rebound of spherical particles. *J. Coll. Int. Sci.*, 1988, **121**(l), 226–239.

31. STENHOUSE, J. I. T. The behaviour of fibrous filters in high inertia systems. *Proc. Filtr. Soc.*, 1972, 426–428.

32. STENHOUSE, J. I. T., BROOM, G. P. and CHARD, N. T. J. High inertia fibrous filtration — optimum conditions. *Proc. Filtr. Soc.*, 1978, March/April, 128–136.

33. WALKENHORST, W. Investigations on the degree of adhesion of dust particles. *Staub Reinhalt. Luft.*, 1974, **34**(5), 149–153 (in English).

34. BARNOCKY, G. and DAVIS, R. H. Electrohydrodynamic collision and rebound of spheres. *Phys. Fluids*, 1988, **31**(6), 1324–1329.

35. WALKENHORST, W. On the effect of the adhesion force on the separation efficiency of dust filters. *Staub Reinhalt. Luft.*, 1972, **32**(6), 30–35.

36. HILLER, R. and LOEFFLER, F. Einfluss von Auftreffgrad und Haftantiel auf die Partikelabscheidung in Faserfiltern (On the degree of impact and the degree of adhesion on particle separation in fibre filters). *Staub Reinhalt. Luft.*, 1980, **40**(9), 405–411 (in German). HSE Translation 9766 (1982).

37. LOEFFLER, F. and UMHAUER, H. An optical method for the determination of particle separation on filter fibres. *Staub Reinhalt. Luft.*, 1971, **31**(3), 9–14 (in English).

38. BROOM, G. P. Adhesion of particles in fibrous air filters. *Proc. Filtr. Soc.*, 1979, Nov/Dec, 661–669.

39. WANG, H. C. and JOHN, W. Comparative bounce properties of particle materials. *Aerosol Sci. Technol.*, 1987, **7**, 285–499.

40. WANG, H. C. and KASPER, G. Filtration efficiency of nanometer size aerosol particles. *J. Aerosol Sci.*, 1991, **22**(1), 31–41.

41. HARPER, W. R. *Contact and Frictional Electrification*, O. U. P., Oxford, 1967.

42. JOHN, W., REISCHL, G. and DEVOR, W. Charge transfer to metal surfaces from bouncing aerosol particles. *J. Aerosol Sci.*, 1980, **11**, 115–138.

43. SMITH, P. A., EAST, G. C., BROWN, R. C. and WAKE, D. Generation of triboelectric charge in textile fibre mixtures and their use as air filters. *J. Electrostatics*, 1988, **21**, 81–98.

44. GILLESPIE, T. and RIDEAL, E. On the adhesion of drops and particles on impact at solid surfaces. 11. *J. Coll. Sci.*, 1955, **10**, 281–298.

45. WAKE, D., GRAY, R. and BROWN, R. C. Assessment of the filtration efficiency of vacuum cleaners used against cotton dust. *Ann. Occ. Hyg.*, 1992, **36**(1), 35–46.

46. PLANTE, W. and JAILLET, J. B. Downstream cleanliness of inorganic filters for ultra high purity gases. In *Advances in Filtration and Separation Technology*, (ed. K. L. RUBOW), Volume 4, pp. 107–111. American Filtration Society, 1991.

47. FISSAN, H., SCHMIDT, W. and HEIDENREICH, E. Quality assurance of gas filtration in microelectronics. *AAAR Conference*, 1991.

48. ACCOMAZZO, A. and GRANT, D. C. Mechanisms and devices for filtration of critical process gases. In *Fluid Filtration: Gas*, (ed. R. R. RABER), Volume 1. American Society for Testing and Materials, Philadelphia, 1986.

49. HOWIE, R. M., ADDISON, J., CHERRIE, J., ROBERTSON, A. and DODGSON, J. Fibre release from filtering facepiece respirators. *Ann. Occ. Hyg.*, 1986, **30**(1), 131–133.

CHAPTER 8

Effects of Loading

Introduction

A captured particle, since it occupies a finite space, becomes part of the filter structure, able to contribute both to pressure drop and to filtration efficiency. Rigorous methods of calculating either of these contributions are limited to systems of relatively simple structure, and so descriptions of the behaviour of clogged filters are incomplete; but it is useful to consider how much can be understood from a simple approach to the problem.

Qualitative description of filter clogging

The simplest conceptual model of a clogged filter is one in which the filter fibres become uniformly thicker as a result of uniformly deposited compact dust. This would mean that the pressure drop of the filter with its altered packing fraction could be calculated using Table 3.1, which suggests that the pressure drop would approximately double as a result of capturing a volume of dust equal to the volume of its fibres. However, the rate of increase of pressure drop is often observed to be very much greater than this; some filters show signs of becoming clogged, with both pressure drop and filtration efficiency increasing, long before the amount of captured dust occupies a large fraction of the interstitial volume. In some cases the pressure drop can be doubled by the capture of dust comprising as little as 1% of the filter volume (1), and so for these filters at least, the simple model is incorrect. In addition the rate of clogging varies with particle size, being greater for finer aerosols, which again invalidates the simple model; and both observations suggest that the captured dust forms a structured deposit within the filter.

In the early stages of filtration, when the number of captured particles is small, the probability of any sort of particle/particle interaction can be neglected. All of the captured particles will be attached to the fibres and, provided that the interception parameter,

N_R, is small, the arguments in Chapter 3 can be used to show that the contribution of the captured particles to the resistance of the filter will be low. Similarly, since an approaching particle is more likely to encounter an uncovered fibre than another particle, the contribution of captured particles to the filtration efficiency will be small. However, as more particles are captured, the likelihood of particle/particle interaction is bound to increase.

Dendrite formation

If a fibre taken from a heavily-loaded filter is examined under a microscope, branched structures made up of deposited particles are seen which, because of their tree-like appearance, are called dendrites. A realistic description of clogging must take into account the dendrites, an example of which is shown in Fig. 8.1; and dendrite development is best studied with reference to each particle capture mechanism in turn.

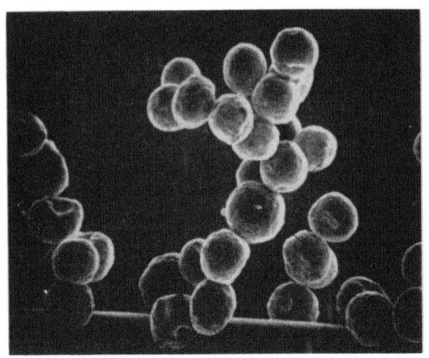

FIG. 8.1   Photograph of a typical dendrite (42). (With permission from Pergamon Press Ltd, Headington Hill Hall, Oxford OX3 0BW, U.K.)

Dendrite formation by interception

Particles captured at a fibre surface by interception are distributed over the leading surface of the fibre; and the normal Stokes flow boundary conditions of zero flow will apply at the captured particle surface as well as at the fibre free surface. A captured particle will be in a region of approximately uniform velocity gradient; and whereas the parts of the particle close to the fibre will perturb the flow only little, those further away may perturb it considerably. The velocity gradient caused by the particle will be largest at those parts of the particle furthest from the fibre surface, and this means, by the arguments developed in the section on interception in Chapter 4, that the most likely further capture site is at the point most removed from the surface of the fibre. This condition will also apply to a particle captured on the original one except that, since it lies in a region where the unperturbed air velocity would be even higher, its own velocity gradient and capture efficiency will be even greater than those of the original particle. The increased velocity gradient means that the drag on an individual dendrite and, therefore, the pressure drop across the filter, increase at a rate that is much faster than linear. A particle captured closer to the forward stagnation point will make a lower contribution to both efficiency and pressure drop, but both of these processes, along with the rate of dendrite growth, would speed up with time.

Dendrite formation by inertial impaction

If the Stokes number of a particle is high its trajectory will cut across fluid streamlines. In the extreme situation, that of infinite Stokes number, its trajectory will be a straight line; and the behaviour of a particle of high but finite Stokes number will tend towards this extreme. Such particles will tend to be preferentially deposited close to the forward stagnation point of the fibre. Approaching particles of high St will have an effective capture area of approximately πd_p^2, where d_p is the mean diameter of the captured particle and the approaching particle.

The capture cross section of the particle considered above is much larger than its cross-sectional area, but it must be remembered that if the captured particle were not there, incident high inertia particles would almost certainly be captured by the fibre, and so the captured particle may not increase the filtration efficiency by very much. The dendrites that grow in this way will tend to project into a region of relatively stagnant flow, and in that form they will not make such a great contribution to the airflow resistance of the filter as will dendrites grown by interception.

Dendrite formation by diffusional deposition

The capture of particles by diffusional deposition even on a single clean fibre involves complicated particle motion, combining the motion of the air with Brownian motion; but a relatively simple solution for dendrite formation can be obtained in the extreme situation, where the diffusivity of the particles is so high that the effects of the convective motion of the air can be neglected. The problem then becomes rather akin to the problem of agglomeration of aerosols, and so is amenable to simple computer modelling, since all that is required is a random walk process. The rate of growth of dendrites will increase as they become larger, because a long dendrite will be reached first by the approaching aerosol; and one effect of the increased capture efficiency will be to make the dendrites more branched. Deposition will take place over the entire fibre surface, and the effect of the dendrites on both capture efficiency and airflow resistance will be higher than that encountered with inertial impaction. A comparison of the effect with that brought about by interception is more difficult.

Dendrite formation by electric forces

The difficulty of making exact predictions of electrostatic effects was shown in Chapter 6, and there is a further complication that electric forces rarely act alone. Particles may, under the influence of electric forces, be captured over the entire fibre surface, and so dendrite formation can be initiated at any point (2) as it can when dendrites are formed by diffusional deposition. Behaviour peculiar to electric forces is the attraction of captured particles as well as airborne particles to the surface of fibres holding a permanent electric charge, which will encourage the collapse of the dendrites and which may reduce the rate at which loading causes the pressure drop to increase (3).

Filters that have an external electric field differ from permanently charged filters in that the former will have a high (applied) field in their inter-fibre spaces. The effect of this is to cause the particles to form long dendrites with relatively few branches, extending in the direction of the field (4, 5); and these have been experimentally demonstrated on isolated fibres with an electric field applied perpendicular to both the fibre axis and the flow direction. This implies that the clogging rate of filters with an external electric field should also depend on the field direction. Further effects of captured aerosol on the behaviour of electrostatic filters will be given later in the chapter.

Illustrations of deposit pattern

As implied above, the form of the deposit will depend on the area of fibre surface over which deposition takes place, a point discussed in Chapters 4 and 6. Typical deposit shapes for each of the three principal mechanical processes are shown in Fig. 8.2 (6). The features described in the section above are clearly visible in the figures, though in practice capture mechanisms may act together, and the deposit itself may change the nature of the capture process.

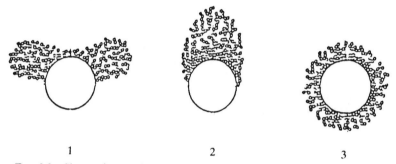

1   2   3

FIG. 8.2 Shape of aerosol deposits on fibres: (1) deposition by interception; (2) deposition by inertial impaction; (3) deposition by diffusion (6). (With permission from Pergamon Press Ltd, Headington Hill Hall, Oxford OX3 0BW, U.K.)

The macroscopic form of deposits can be included in the calculation of flow pattern and filter resistance as carried out in Chapter 3. The simplest approximations to the fibre shape in the cases of interception and impaction deposition are ellipses, the major axis of the ellipse being perpendicular to the flow in the first case and parallel to it in the second. The behaviour of such fibres can be described by complex variable theory (7) or by many-fibre theory; and the calculated effect of this on pressure drop has been discussed in Chapter 3 and on single fibre efficiency in Chapter 4.

The influence of aerosol properties on clogging rate

Some early but still unchallenged observations of the effect of aerosol and filter properties on clogging rate (1) showed that very coarse dust does not penetrate into a filter but forms a cake, contributing little to either the resistance or the efficiency of the filter. Dust fine enough to penetrate has a very different effect, causing both resistance and efficiency to increase. Structures built up of fine aerosols would tend to resemble fine fibres; and previous work, described in Chapter 3, has shown that these have an effect on the airflow resistance of a filter made from polydisperse fibres that is large in comparison with the fraction of the filter volume that they occupy (8). Acicular particles and smoke agglomerates are both highly clogging, the former because they resemble fine fibres and the latter because they are rather like established dendrites captured during a single process. Liquid aerosols tend to reduce filter efficiency and pressure drop, though exceptions to this behaviour will be discussed later in this chapter.

Macroscopic theory of filter clogging

Filters with an open structure or a low packing fraction tend to be resistant to clogging, the reason being that dendrites would have to grow to a considerable size before they would seriously limit the pathways that the air could follow when traversing the filter. This effect has been shown clearly in practice (9) with two filters of comparable efficiency but with different packing fractions. One, with a packing fraction of 0.07, had approximately three times the dust-holding capacity of the other, with a packing fraction of 0.32.

Layer efficiency also influences clogging rate by affecting the deposition pattern within the filter. Consider for example two filters of identical thickness, one of which captures 99% of the incident aerosol and the other 99.9%. Although the mass of captured material is almost the same for the two filters, the higher layer efficiency of the second will mean that the aerosol is captured in a smaller volume, as Fig. 8.3 indicates. The most highly loaded part of the filter becomes the most efficient since clogging increases efficiency, and this part also receives the highest challenge of aerosol, resulting in a greater tendency of the filter to clog at the leading surface.

The extreme case of clogging of a high efficiency filter spans the gap between depth filtration and surface filtration. As stated in Chapter 1, surface filtration is characteristic of fabric filters. The capture efficiency of the first layer of fibres in these filters, which are used against coarse dusts, is high. The dendrites formed here increase the efficiency still further, with the result that they rapidly coalesce to a single porous mass, initiating the formation of a dust cake. Depth filtration in such filters occurs only to a small extent. Even in fibrous filters, however, deposition is most effective in the initial layers, and as the deposition rate proceeds there is a tendency to convert to surface filtration. The development of this mode of action is often the final behaviour regime of a clogged fibrous filter.

The correlation between clogging rate and filtration efficiency is taken into account in what is probably the most popular means of controlling the clogging of a filter, the use of layered structures. A layered filter can be made from composite materials such that the least

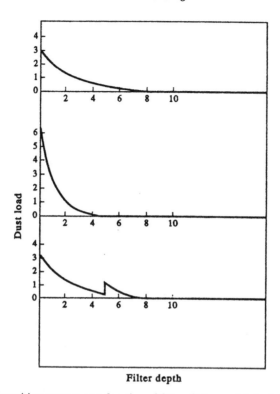

Filter depth

Fig. 8.3 Deposition pattern as a function of filter efficiency: (a) single layer 99% efficient filter; (b) single layer 99.9% efficient filter; (c) double layer 99.9% efficient filter.

efficient is first challenged by the aerosol; or it may be made by using material of a common fibre size, but with packing fraction varying so that the more open filter is challenged first. An illustration of this effect, for a hypothetical two-layer filter with the same efficiency as filter 2, is shown in Fig. 8.3 as filter No. 3. The high efficiency component, though more susceptible to clogging, receives a low dust load.

This concept can be put on a more rigorous basis (10, 11) using a simplified form of the mass balance equation (12, 13) which relates a variable layer efficiency to the rate of change of concentration as the aerosol passes through the filter.

$$\alpha(x) = \frac{-dN(x)}{dx} \cdot \frac{1}{N(x)} \tag{8.1}$$

It is a relatively simple matter to show that a uniform deposit of dust can be achieved by a filter of depth h with penetration P if,

$$N(x) = N(o)\left[1 - \frac{(1-P)x}{h}\right] \qquad (8.2)$$

and so,

$$\alpha(x) = \frac{(1-P)}{h-(1-P)x}. \qquad (8.3)$$

A problem with such a structure is that the layer efficiency of the downstream parts of the filter would be extremely high, as substitution of $x = h$ in equation 8.3 shows; and this is likely to mean excessive clogging at this point. A minimum-clogging solution is likely to lie between this structure and a homogeneous filter.

In addition the polydisperse nature of real dust would have to be accounted for, along with the manner that the size distribution of the partly filtered aerosol, and its tendency to clog the filter, would change with both depth and time. These complications mean that the design of reduced-clogging filters must be specific to conditions.

Airflow pattern and drag force acting on particle complexes

The behaviour of filters will depend on the airflow around dendrites, which will resemble the airflow around particle clusters; and a first step towards this is a study of the flow around a single particle, assumed, for simplicity, to be spherical.

Theory of airflow around an isolated sphere

The viscous drag acting on an isolated sphere in Stokes flow and the pattern of flow around it form part of the fundamental theory of fluids, and the theory will be summarised here, for it will aid the understanding of more complicated models later in the chapter.

Both the flow pattern and the drag follow from a solution of equation 3.24, the biharmonic equation, in the appropriate coordinates and with the appropriate boundary conditions. A method of approaching this that will be referred to in further studies is the use of a singularity known as a stokeslet (14).

It has been shown that two-dimensional flow is most easily described by means of the stream function, ψ. Similarly, when the flow pattern has circular symmetry about an axis, so that the air velocity is non-vanishing in the radial and angular directions only, as shown in Fig. 8.4, these components of fluid velocity are most easily described by means of a scalar, ϕ, to which they are related by,

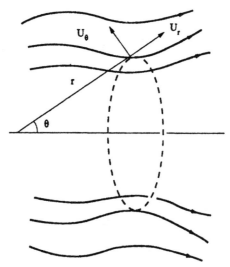

FIG. 8.4 Illustration of radial stream function, related to air velocity by equations 8.4 and 8.5.

$$U_r = \frac{1}{r^2 \sin\theta} \frac{\partial\phi}{\partial\theta} \tag{8.4}$$

$$U_\theta = -\frac{1}{r \sin\theta} \frac{\partial\phi}{\partial r}. \tag{8.5}$$

With this notation, a stokeslet is given by,

$$\phi_s = \frac{Fr \sin^2\theta}{8\pi\eta} \tag{8.6}$$

where F is the force acting by viscous drag. The stokeslet alone cannot satisfy the boundary conditions of zero U_r and U_θ at the surface of a sphere in uniform flow, but it forms part of the appropriate stream function, which is, for a sphere of radius R, in a flow of uniform velocity U,

$$\phi = UR^2 \left(\frac{r^2}{2R^2} - \frac{3r}{4R} + \frac{R}{4r} \right) \sin^2\theta. \tag{8.7}$$

The drag on the sphere, due entirely to the stokeslet, is,

$$F = 6\pi\eta UR \tag{8.8}$$

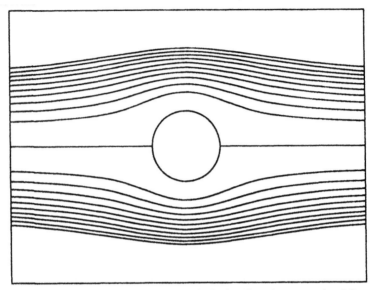

FIG. 8.5   Streamlines around a sphere in Stokes flow, from equation 8.7.

which is the classical value. The streamlines for flow around a sphere are sketched in Fig. 8.5, corresponding with equation 8.7.

Aerodynamic slip

The theory above applies to the continuum regime; but like the theory of flow around a fibre, it can be generalised to take account of aerodynamic slip, by inclusion of the appropriate dimensionless parameter, the Knudsen number Kn, which causes equation 8.8 to be modified to (15),

$$F = \frac{6\pi\eta UR}{1 + Kn\left[A + Q\exp-\left(\dfrac{B}{Kn}\right)\right]} \tag{8.9}$$

where the constants A, Q and B have the values 1.246, 0.42 and 0.87 respectively, as in equation 4.35.

Drag force acting on agglomerates

The next step is the extension of the drag force from a single sphere to that of an assembly of spheres, simulating a dendrite. An analytical treatment of the flow around two touching identical spheres, a doublet

(16), is much more difficult than that for a single sphere. As the number of spheres increases, exact analytical solutions become more difficult still, and the flow pattern can only be reasonably imagined. On the other hand the viscous drag acting can readily be obtained by measuring the settling velocity of large but geometrically similar models of clusters through viscous fluids. Such experiments have been carried out (17) on clusters and on chains of spheres up to five particles in length.

Even a doublet has lost the spherical symmetry of the single sphere; the simple formula whereby the force and flow velocity are related by a scalar quantity no longer holds. The quantity relating them becomes a tensor, and the drag becomes orientation dependent as described for fibres in Chapter 4. However, in the case of linear chain, which is the simplest approximation to the dendrite structure, the drag at any orientation can be expressed as a simple combination of the drag when the flow is parallel to the axis of the chain and the (invariably larger) drag when it is perpendicular.

If each sphere in a doublet were to act independently, the drag would be twice that on a single sphere. In fact it is more than that on a single sphere but less than twice this value, increasing monotonically with length, but with a dependence on length that becomes weaker as the chain becomes longer. Theoretical expressions for the drag acting on prolate spheroids have been fitted (18) to the above experimental results, for chains with lengths of up to five particles (17). A good fit is obtained with two adjustable parameters, which are then used in extrapolations to longer chains. The results of these calculations are given in Table 8.1. Linear agglomerates are the best simple approximations to dendrites, but experimental measurement of the aerodynamic properties of agglomerates of more complicated geometry have also been made (17, 19).

The factors in Table 8.1 are the quotients of the drag acting on the particle chain and that acting at the same velocity on a single isolated constituent particle. This parameter must be distinguished from the dynamic shape factor (20), which is the quotient of the drag acting on a structure and that acting at the same velocity on a sphere of the same volume.

Aerodynamic slip

Aerodynamic slip is important in the description of the behaviour of chains of small particles and therefore of dendrites, but its extension beyond a single particle requires approximations. The value of R in equation 8.9 for slip around a spherical body must be modified to the effective radius of the chains, which depends upon the orientation of the

TABLE 8.1

Drag on straight chains of spheres in uniform flow, expressed as a multiple of the drag on a single sphere

No of spheres	Continuum flow		Kn=0.2	
	Parallel	Perpendicular	Parallel	Perpendicular
1	1.00	1.00	0.80	0.80
2	1.27	1.44	1.07	1.22
3	1.53	1.82	1.30	1.58
4	1.77	2.20	1.52	1.93
5	1.98	2.52	1.73	2.23
6	2.19	2.85	1.93	2.55
8	2.59	3.47	2.30	3.13
10	2.98	4.07	2.66	3.70

chains with respect to the flow (21), but which is smaller than the equivalent volume sphere radius (18). Table 8.1 shows the value of the drag force for chains of spheres at a value of $Kn=0.2$.

Drag force on a single particle attached to a collector

The simplest method of calculating the drag on an attached particle is to approximate it by the drag, given by equation 8.8 (22), that the particle would experience in a uniform flow field with the velocity of the unperturbed flow at the particle's centre calculated from any theory of flow. A refinement to the theory is the inclusion of Faxen's correction (23), replacing equation 8.8 by,

$$F=6\pi\eta R\left(U+\frac{R^2}{6}\nabla^2 U\right) \qquad (8.10)$$

which takes some account of the non-uniformity of the flow by introducing a term proportional to the second derivative of the velocity. The first derivative, as shown in Chapter 4, will tend to rotate the particle but it will not exert any net translational force. This refinement can, however, produce results that are further from what is observed than are those arising from the simpler expression.

A more complete approach (24) results in a correction factor, f, to the expression for the drag acting on a spherical body in the vicinity of a plane surface in shear flow.

$$F=f6\pi\eta UR \qquad (8.11)$$

For the case where the particle is in contact with the surface the correction factor is calculated to be 1.70, which is in good agreement with the experimentally observed value (25).

Drag force on a dendrite attached to a collector

The drag force acting on a dendrite varies not only with its length, but also with its orientation. In the extreme cases of a dendrite at the forward stagnation point and one at the widest part of the fibre, the unperturbed velocity components acting will be respectively the radial and the tangential velocity around the fibre. The radial velocity varies approximately quadratically with distance from the fibre surface, and the tangential, which is stronger, approximately linearly. The effect is augmented by the fact that the drag acting along the axis of a chain is weaker, and so we would predict a much smaller effect on the former dendrite, especially when it is short, as outlined above in the simple description of the effect of particles captured at different positions on the fibre surface.

The theory previously developed, in which a chain of particles is approximated by a prolate spheroid, can be extended to dendrites in shear flow (26). The chosen approximation is a half-spheroid, with its cut surface in contact with the fibre, at the point of zero flow, the form of the spheroid being such as to have the same height and total surface area as the dendrite.

A further method, suitable for calculating the drag on chains of spherical particles (27) follows from slender body theory, in which a body with a large aspect ratio is approximated by a series of stokeslets like those described above, paired with further singularities, potential dipoles, necessary to satisfy the boundary conditions. The approximation gives an exact flow pattern at large distances from the body. The positions of the singularities can be made to coincide with those of the particles making up the dendrite. In addition each singularity must be taken with its image in the surface of the collector, which should, ideally, preserve the boundary conditions of zero velocity at the collector surface. The example cited contains calculations confined to a spherical collector, which allows much easier calculation of images than does a cylindrical collector. The results are, therefore, not directly applicable to fibrous filters, though the principles emerging will be. Moreover, the number of singularities being relatively small, it is possible to satisfy the zero velocity conditions only at an equal finite number of points. These are chosen to be the forward stagnation points of the spheres making up the dendrite, and so the solution must be regarded as a step rather than a final answer. However, the theory does contain the important feature that it accounts at least in part for the effects of the collected particles on the airflow pattern.

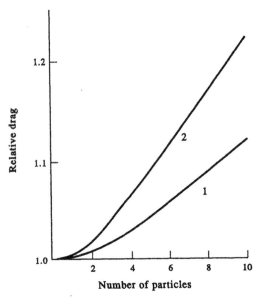

FIG. 8.6 Relative drag on dendrites as a function of number of particles contained. The function is discrete, but it is illustrated in continuous form for convenience. The ordinate is the drag on the (spherical) collector plus dendrite divided by the drag on the collector alone: (1) quotient of particle diameter and collector diameter = 0.1; (2) quotient of particle diameter and collector diameter = 0.2 (27). (© 1988, Elsevier Science Publishing Co. Inc.)

Figure 8.6 shows the variation of dendrite drag with length calculated by this method, normalised to the drag acting on the collector alone, and it can be seen that the increase in drag is roughly linear. The effect of angle of orientation on the drag of a five particle dendrite, calculated by this method, is shown in Fig. 8.7.

Theoretical description of clogging

It was shown, in Chapter 3, that theoretical models of airflow through clean filters successfully predicted what was observed with real filters. The reasons for this are that the fundamental assumption made in their derivation, the Stokes flow approximation, is a good one, and that the principal discrepancy between model filters and real filters, the geometric simplicity of the former and complexity of the latter, has a relatively small effect. The pressure drop of filters of common packing fraction varies relatively weakly with structural differences.

The agreement between theory and experiment with respect to clogging is much less satisfactory. Clogging is a time-dependent process, and even the detailed ordering of events is critical; the effect of

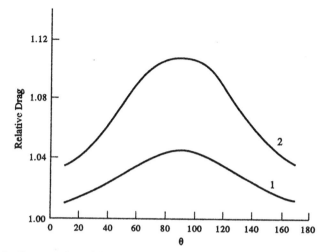

FIG. 8.7 Drag on 5-particle dendrites as a function of angle of deposition. The ordinate is the drag on the (spherical) collector plus dendrite divided by the drag on the collector alone. (1) and (2) are calculations by different methods (27). (© 1988, Elsevier Science Publishing Co. Inc.)

two particles approaching a fibre will depend on which approaches first, and this may even decide whether one particle is captured or both. In principle one would expect ensemble averaging to remove this complication, but even an ensemble average will depend on the microscopic structure of the filter. Flow and particle capture by clean filters can be well-modelled as a two-dimensional process, but if the simplest capture event of all occurs, the capture of a spherical particle, the symmetry of the model is destroyed. The growing dendrite will significantly alter the flow pattern in its own immediate vicinity, and as the population of dendrites grows they will influence each other. The evolution of dendrites depends critically upon the position of the initial captured particle, and this may depend on the flexibility of the fibre. Particle bounce at the fibre surface will influence growth, as will the bouncing of particles from established dendrites. The energy transfer during capture may influence the shape of the dendrite, if it is not rigid, as may static forces acting among particles or between a particle and the fibre.

The complete problem is intractable, and no progress in a simulation can be made without considerable approximation. It is not possible completely to separate theory from experiment, since authors often put the two side by side; and in the work that follows most of the theory deals with isolated fibres, single fibres and simple model filters of high symmetry.

Calculation of dendrite formation

Calculation of the actual process of dendrite growth must take account of the efficiency of dendrites in particle capture, since this is the way that they grow. However, a general caution must be made against giving a model credence simply because it predicts the formation of dendrites. Almost any model will do this, and so qualitative agreement with experimental observation is easy to obtain. Quantitative agreement is much more difficult.

The simplest theory (28) involves a series of coupled equations in which the rate of change in the population of dendrites of a particular size is related to that of dendrites with one particle fewer. The equations can be solved if the approximation is made that the probability of a particle capturing a further particle is independent of the site of the particle. In this situation the total number of particles, m_{tot}, captured within time t, by an originally clean fibre of length L, is given by,

$$m_{tot} = \frac{4E_s \, d_f L}{\pi E_{sp} \, d_p^2} \left[\exp(\pi \, d_p^2 E_{sp} NUt/4) - 1\right] \qquad (8.12)$$

where E_s is the capture efficiency for the clean fibre, E_{sp} is the capture efficiency for a particle, and NU is the rate at which particles approach the fibre. A complete theory would require the capture efficiency of dendrites of all sizes and shapes to be known; but the simplified equation shows that particle capture becomes much more likely as loading proceeds.

A more detailed but still approximate mathematical theory of dendrite formation (29, 30) describes the dynamics of capture of particles by dendrites, dendrite morphology, and the changes in pressure drop and penetration. The particles approaching the fibre are assumed to be monodisperse, and dendrite structures are idealised to a straight chain structure, as shown in Fig. 8.8, which is a simple layered

FIG. 8.8   Real and idealised dendrite (29). (With permission from Pergamon Press Ltd, Headington Hill Hall, Oxford OX3 0BW, U.K.)

form enabling the particles to be classified into a discrete number of sets according to the layer that each occupies. The dendrites are assumed to be rigid and the particles forming them are assumed to have no effect on the flow field. An expression for m_k, the number of particles in the kth layer, as a function of time, t, is written down,

$$m_k = \rho_m^{k-1}\left[1 - \sum_{j=1}^{k-1}\left\{\prod_{\substack{1 \le i \le k-1 \\ i \ne j}} \frac{\phi_i}{(\phi_i - \phi_j)}\right\} \exp\left(-\frac{\phi_j t}{\rho_m}\right)\right]. \quad (8.13)$$

ρ_m is the maximum number of particles in the $k+1$st layer that can be attached to the same particle in the kth layer, and ϕ_k is the rate constant for particle deposition on a particle in the kth layer of the dendrite. m_k is a discrete variable, whereas the right hand side of equation 8.13 is a continuous variable. A particle must be assumed to have arrived when m_k reaches an integral value.

The expression above is general, but the values of the rate constants are not. For an unperturbed Kuwabara flow field, with the motion of the particles confined to a single plane perpendicular to the axis of the fibre, the expression for the rate constant appropriate to capture by interception only, is,

$$\phi_k = \frac{\pi\, d_p^2 UN}{4Ku}\left[2 \ln(1 + 2kN_R) + 1 + c - \left(1 - \frac{c}{2}\right)(1 + 2kN_R)^{-2}\right.$$

$$\left. - \frac{3c}{2}(1 + 2kN_R)^2\right] \sin\theta_0 \quad (8.14)$$

where N is the concentration of particles in the incident air and θ_0 is the angular position of the dendrite.

According to this model, the form of the dendrites is a function of the product of the incoming aerosol concentration and the lapsed time. The filtration efficiency can be calculated by differentiating equation 8.13 and summing over the dendrite layers and the angles of deposition. The pressure drop can be calculated by applying the previous theory of the drag produced by the air on the components of the dendrites, and summing this over all of the deposited particles.

In later work account is taken of the modification to the flow caused by the particle in the dendrite that was being targeted by the approaching airborne particle (31) but not by the other particles in the dendrite. A feature that emerges in this calculation is that the maximum growth rate is experienced by dendrites with a point of attachment to the fibre about $\pi/3$ from the forward stagnation point.

Diffusional deposition is included in the theory (32) by adding to the working a numerical solution of the diffusion/transport equation within the model, an approach that is consistent with the methods used in the model formulation. Dendrite initiation can occur at any point on the fibre surface, though the probability decreases monotonically from the forward to the rear stagnation point. However, just as in the case of interception capture, the maximum rate of growth of dendrites when diffusion takes place is observed at some angle ($\pi/4$ in the case chosen) from the forward stagnation point.

Numerical simulation of dendrite growth

In the approach used above (29) the dendrite formation was calculated as a function of the angular coordinate of the deposition site of the first captured particle, and was essentially a deterministic process. An alternative approach is to use a Monte-Carlo calculation, in which particles are introduced into the flow at random positions upstream from the fibre, some being collected and others passing through. The collected particles are allowed to act as collectors of subsequently introduced particles, and so dendrites may build up. A series of calculations may be carried out and an average taken.

The method predicts, as would be expected, that the dendrites tend to grow on the front face of the fibres when the Stokes number is high, and at the edges when it is low (33). Shadowing of regions on a fibre surface immediately downstream of a dendrite is also observed. For capture by interception with $N_R = 0.05$ and a deposition angle $10°$ away from the forward stagnation point (34), a two-particle dendrite will shield the entire half of a collector whereas for inertial capture with a Stokes number of 3.0, about thirty particles would be needed to do this. Although the collector in the latter example was a sphere, the results will apply in principle to a fibre.

In further work (35), applied to cylindrical collectors, a direct comparison between the theoretical predictions and experimental observations was made. The Oseen flow field for an isolated fibre was used, and the particles were introduced into the flow at random. The deposited particles were assumed to have no influence on the flow field, and contact was assumed to be sufficient to ensure capture. The mean of 10 results was taken as a usable average. The extent of agreement between theory and experimental values was variable, an example of good agreement being shown in Fig. 8.9. The straight line indicates the calculated behaviour of a clean fibre, the curved line shows the calculated behaviour of the fibre plus dendrite, and the points show the results of experiments.

Other authors have used the Monte-Carlo method, with the same

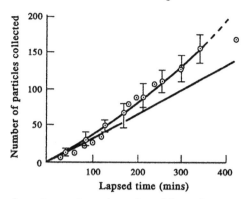

FIG. 8.9 Time dependence of number of particles collected by a fibre; ⊙ – experiment; —— theory, both accounting for and neglecting the effect of dendrites (35). (By permission of Gordon and Breach, Science Publishers.)

rigid dendrite approximation, and with Kuwabara flow (36). Figure 8.10 shows the fractional increase in single fibre efficiency as a function of dust load, calculated by this approach, and the contrast between the high increase at low Stokes number and the low increase at high is considerable. The authors found their results in basic agreement with those of experiment but, as is often the case, theoretical predictions showed a stronger dependence on the relevant parameters than did experimental results.

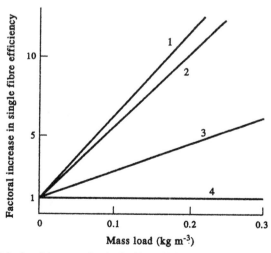

FIG. 8.10 Calculated increase in single fibre efficiency as a function of deposited mass, for various Stokes numbers: (1) $St = 0.0$; (2) $St = 0.6$; (3) $St = 1.0$; $St = 2.0$ (36). (With permission from Pergamon Press Ltd, Headington Hill Hall, Oxford OX3 0BW, U.K.)

Like the deterministic model this Monte-Carlo approach predicts that total particle deposition would be greatest at a finite angle from the forward stagnation point. For pure interception this angle is approximately $\pi/6$, and it decreases as the Stokes number increases. The reason for this can be readily understood for the case of pure interception. It was shown in Chapter 4 that deposition of particles on a clean fibre, by this process, was greatest at the forward stagnation point, falling to zero at an angle of $\pi/2$. However, it was shown earlier in this chapter that a deposited particle makes the greatest contribution to increased filtration efficiency when it is deposited at an angle of $\pi/2$, and that this falls steadily with decreasing angle, to reach a minimum at the forward stagnation point. The combined effect of these two processes is to give a maximum in dendrite growth at some angle between these two limits.

The Monte-Carlo approach has been used in simulations of dendrite formation by diffusional deposition (37) by adding a random step of appropriate size to the motion of the particle during the trajectory calculation. The rate at which the single fibre efficiency increases is greatest when Peclet number is highest (capture efficiency by diffusional deposition is lowest) and whereas for high values of the Peclet number particle deposition is highest at a finite angle from the forward stagnation point, at low Peclet numbers this angle drops to zero.

Further numerical work has been carried out by a number of workers (38, 39, 40), using spherical polydisperse simulated aerosol particles. Some authors have also taken some account of the electric force between the fibre and the aerosol. An example of the results obtained is sketched in Fig. 8.11, though it is difficult to generalise.

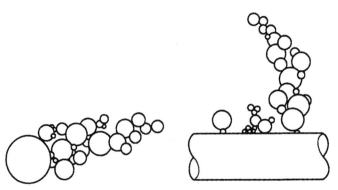

Fig. 8.11 Simulation of dendrite grown from polydisperse particles on an electrically charged fibre (39). (With permission from Pergamon Press Ltd, Headington Hill Hall, Oxford OX3 0BW, U.K.)

Experimental observation of dendrite structure on a single fibre

Experimental observations have been made (41) of the capture efficiency of single fibres of 8 μm diameter exposed to monodisperse aerosols at a range of velocities and therefore, because of the effects of inertial impaction, with different collection efficiencies. In the clean state the single fibre efficiencies of the fibres varied by a factor of more than 20, but after exposure to a sufficient number of particles to cause a 250 μm length of the fibre to collect 100–200 particles the single fibre efficiencies varied by only a factor of 2. The results indicate the importance of particle–particle collision as a capture process, and demonstrate that inertial effects are less important than interception in clogging.

Dendrites built up from 3.3 μm methylene blue particles deposited on a 25 μm fibre have been examined by optical and electron microscopy (42, 43), and the number of particles in each dendrite particle layer were counted and compared with theoretical calculations (employing a Kuwabara flow field, rather than an isolated fibre flow field). The observed angular distribution of dendrites was found to vary more weakly than the calculated distribution; and dendrites were found at angles greater than $\pi/2$ from the forward stagnation point, which simple theory would contradict. It was suggested that the captured particles tended to skid on the fibre surface, lubricated by a cushion of air before impact, as a result of the molecular nature of the air; though other workers (44) have disputed this and attributed the effect to particle bounce (*vid sup*). A further possibility is that a particle may roll, under the influence of the airflow, especially if its initial area of contact with the fibre is small.

Experimental observations on model and real fibrous filters

There are no generally applicable equations describing the pressure drop or penetration of clogged filters, but the following empirical relationship has been obtained for penetration (45):

$$\frac{P(M_{\rm L})}{P(o)} = \exp(-\alpha M_{\rm L}^{2/3}) \qquad (8.15)$$

where α is a constant that depends on the filters and the aerosol. Other workers found that a linear team in $M_{\rm L}$ gave a better fit than a 2/3 power (13, 46).

The following empirical relationship between the change in pressure drop and that in aerosol penetration for a clogged filter, was observed to fit experimental data (47),

$$\ln\left[\frac{P(t)}{P(o)}\right] = -\gamma\left(\sqrt{\frac{\Delta p(t)}{\Delta p(o)}} - 1\right). \tag{8.16}$$

The subscripts t and 0 apply to filters in the loaded and clean condition respectively. The pressure drop increased exponentially with dust load, and both this and the relationship in equation 8.16 were observed in many filters, but not all; though the coefficient of the exponential and the parameter γ differed among filters.

It is quite likely that other simple expressions could be fitted to other sets of experimental data, to be valid over a limited range of variables. However, the range of validity can be better defined if specific attention is paid to the capture mechanisms in operation.

Behaviour related to capture mechanisms

In experiments on real fibrous filters (48, 49) in the interception/inertia regime, the change in single-fibre efficiency was,

$$\frac{E_S(M)}{E_S(o)} = 1 + \lambda M \tag{8.17}$$

where M is the mass of deposit per unit volume of filter material. (A similar functional form was found in simulations of capture by diffusional deposition (37), and since it is a basic linear approximation, it may apply quite widely.) The efficiency enhancing factor, λ, is specific to the actual filter used, but the authors found that it was greater at low loading velocities, varying between 10 and 0.1 m^3/kg as the Stokes number varied from 0.05 to 1.0, and that it was relatively insensitive to the interception parameter. When diffusional deposition was dominant, microscopic examination showed that dendrite growth was much more rapid; and it was also observed that the dust load required to double the resistance was only about one tenth of that during inertial deposition. Other authors (50), loading model mesh filters of different grades, observed that the behaviour of the filtration efficiency was reasonably well described by the following expression for the mass of aerosol per unit area of filter measured in kg/m^2 which, when deposited, caused a doubling of the single fibre efficiency,

$$M_0 = AE_S U^{0.25} \tag{8.18}$$

where E_s is the single fibre efficiency of a clean fibre, U is the approach velocity of the air in cm s^{-1}, and A is an index which varies between 2.17×10^{-3} for a 200 mesh and 1.1×10^{-3} for a 500 mesh. Though a similar pattern might be followed by a range of filters, the values of the coefficients are system-specific.

Experimental measurements of the pressure drop increase in filters clogged by aerosols of different size, and therefore acting by different mechanisms, are shown in Fig. 8.12 (51, 52). As the theory would suggest, the clogging rate in the inertial regime is lowest, and that in the diffusional regime the highest, with interception between.

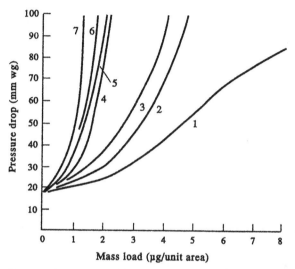

FIG. 8.12   Pressure drop across a filter loaded at 0.1 ms^{-1} as a function of mass of dust held in the form of particles of different sizes: (1) 0.6 μm; (2) 0.4 μm; (3) 0.3 μm; (4) 0.2 μm; (5) 0.15 μm; (6) 0.1 μm; (7) 0.07 μm (51). (By permission of the Aerosol Society of Great Britain.)

HEPA filters have been loaded with aerosols (53) and found to behave in such a way that the rate of clogging, measured as pressure drop against mass of deposit captured, did not vary with the air velocity at which loading took place. This could be explained if the principal deposition process is interception, since this is a velocity independent process, and provided that interception by the dendrite is the principal process by which the dendrite acted, the velocity independence would be preserved here as well.

Effect of filter properties on the pattern of clogging

A comparison has been carried out between model filters and real filters of different packing fractions (49). As would be expected the increase in pressure drop with load was greater for the filter with the largest packing fraction.

In one set of experiments meshes of wire with diameters between 25 and 50 μm in stacks of 1 to 16, with inter-layer spacings of 2 mm, were challenged with 1.1 μm monodisperse aerosols of stearic acid (50). The Stokes number of the incident particles was varied between about 0.001 and 0.3, and the variation with the load of the collection efficiency of a single layer loaded at different velocities was observed.

The collection efficiency of a single layer increased gradually with dust load until a value of about 0.1 was reached, after which a catastrophic increase in both filtration efficiency and pressure drop occurred. The load at which the sudden increase took place was lower for lower loading velocities, and with these the collection efficiency reached 1.0. At higher velocities the onset was delayed and, instead of reaching 1.0, the collection efficiency reached a maximum and then dropped again, probably because of particle bounce. The pressure drop followed a similar pattern for the low velocity loads, but in situations where the collection efficiency did not reach 1.0, it did not vary in a systematic way.

A similar pattern, of pressure drop increasing gradually with load for a relatively long period, and then suffering a catastrophic increase, was observed over a time period of months, in the case of air-conditioning filters exposed to atmospheric aerosol (54).

Relatively low efficiency filters (52) with a fibre diameter of 4.7 μm, loaded with aerosols of diameter 1.16 μm, gave plots of penetration and pressure drop as a function of load that had the same shape irrespective of the thickness of filter used in the test, differing only by a common ordinate. This indicated that clogging took place essentially in the first layer of filter used, the rest of the composite receiving very little load. Kozeny Carman theory was used to study the structure of the deposit, showing that it had a similar packing fraction to that of a deposit formed during a high inertia filtration simulation. The authors consistently found a discontinuity of gradient in the plot of pressure drop against dust load, as illustrated in Fig. 8.13, which they termed the "clogging point" and which was associated with a change in the mode of operation from depth filtration to surface filtration. A similar pattern was observed in earlier work (55).

HEPA filters with fibres of the order of 1 micrometre in diameter have been loaded with dust that formed a cake on the filter surface (56). The cake is rather like a bed with grain size equal to that of the challenging aerosol particles. The authors observed, as have others (57), that the decrease in penetration and the increase in pressure drop both tended to be greater for the finer particles. When the size of the particles was kept constant but the velocity of filtration varied, it was

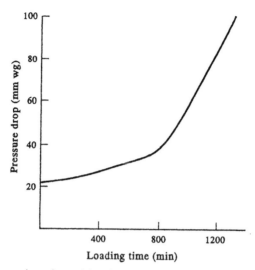

FIG. 8.13 Illustration of transition from depth to surface filtration. The sudden change in gradient of the graph is clear (52). (With permission from Pergamon Press Ltd, Headington Hill Hall, Oxford OX3 0BW, U.K.)

found at lower filtration velocities, both that the initial penetration was lower, indicating diffusional capture, and that the gradient of penetration, on a logarithmic scale, against mass load on a linear, was more steeply negative.

Particle bounce from clogged filters

The possibility of particle contact with a collector, without capture, is significant for clogged filters as well as for clean ones and, as with the latter, its likelihood is greater under conditions of high particle inertia (58). This was confirmed by theory and experiment for a model filter consisting of a single layer of 25 μm metal fibres wound with a spacing of 50 μm to produce a sheet of parallel fibres (59), exposed to 2.02 μm monodisperse particles of poly vinyl toluene latex.

The theoretical model used for the flow was that appropriate to an isolated row of fibres (60), based on the complex variable approach, as described in Chapter 3. Trajectory calculations were carried out numerically, and capture was assumed to take place by interception, modified by inertial effects. The flow field was assumed not to be perturbed by the existence of the dendrites; but bounce on particle–fibre or particle–particle contact was modelled using the theory described in Chapter 7, accounting for van der Waals forces, and particle deformation, giving rise to a calculated value of critical impact velocity. Several simulations were carried out for each set of conditions because the dendrite development was modelled as a stochastic process.

In the experiments filtration efficiency was observed to increase from about 1% to almost 100% and so clogging was severe. A similar order of increase in the pressure drop was observed.

The extreme case, of filter clogging with particle bounce in the limit of high Stokes number, has been studied by numerical simulation (61), using theory previously described (62). Clogging in high inertia filtration has also been studied experimentally (63), giving results, illustrated in Fig. 8.14. Penetration fell with diameter due to inertial impaction, which was insensitive to dust load, followed by an increase, due to particle bounce, which became less severe as loading proceeded. The results indicated, as discussed in Chapter 7, that a layer of collected dust particles, particularly larger particles, gives rise to less bouncing than a clean fibre.

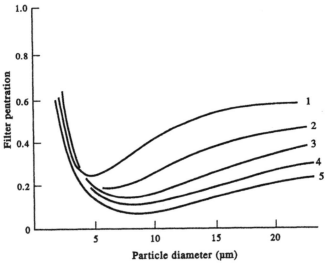

Fig. 8.14 Effect of loading, at air velocity 1 ms^{-1}, on aerosol penetration: (1) dust load 0.01–0.1 g; (2) dust load 0.25 g; (3) dust load 0.5 g; (4) dust load 0.75 g; (5) dust load 1.0 g. At higher dust loads the penetration increased, but in an irregular way (63). (By permission of the American Industrial Hygiene Association.)

Fractal structure of particle deposits

Just as the position of deposition of particles captured by clean fibres varies according to the capture process operating, so will that of particles contributing to dendrites or other structured deposits. This will affect the macroscopic shape taken by the deposit (6) as illustrated in Fig. 8.2. A parameter that describes the shape on a smaller scale would be useful, and a contender for this is the fractal dimension of the dendrites.

A brief account of fractal dimension will be given here, though detailed accounts are available (64, 65). If a deposit were built up in a manner in which its (three-dimensional) shape did not alter with its size, then the mass of the deposit or the number of particles contained would vary as the cube of the linear dimensions of the aggregate; its dimensionality would be three. If the deposit formed in a plane, with constant shape but varying size, the dimensionality would be two; and if it formed as a chain its dimensionality would be one. Fractal geometry allows non-integral dimensionality to be ascribed to systems in which the number of particles varies as a non-integral power of the linear dimension. For instance the particle complex shown in Fig. 8.15, in which a circle of radius a, drawn from the centre, contains a number of particles proportional to $a^{3/2}$, has a fractal dimension of 1.5 (6).

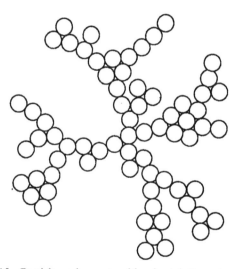

FIG. 8.15   Particle agglomerate with a fractal dimension of 1.5.

The physical property that can be most simply related to fractal dimension is radius of gyration, which is proportional to the number of particles in a cluster raised to the power of the reciprocal of the fractal dimension (66). However, the drag forces acting on a dendrite will depend not on its radius of gyration, but on its hydrodynamic radius (the radius of the sphere that gives the same streamline pattern at large distances) which depends on its size and its shape and, therefore, its the fractal dimension. Intuitively, one would expect a cluster with a low fractal dimension to experience a high drag.

An account has been given of the drag experienced by assemblies of particles that have a well-defined fractal dimension (67). The results are not directly applicable to dendrites since the latter are subjected to a flow gradient; and the authors point out that the fractal dimension alone gives an incomplete description of the aggregate. The results are expressed in terms of the quotient of the radius of gyration and the hydrodynamic radius. For a dimension of unity this quotient varies according to the logarithm of the number of particles in the cluster (which equals the aspect ratio of the cluster). For a dimension of 3, the quotient is a constant. For a fractal dimension greater than about 1.3 it rapidly reaches a value that does not change with the number of particles in the cluster.

A deposit may be characterised by the fractal dimension of its outline, expressed as the variation of perceived length with increment size. The fractal dimension of dendrites produced by 4 μm diameter particles, experimentally deposited, probably by inertial impaction, at a filtration velocity of 30 cm sec^{-1} on to a 25 μm diameter fibre, have been measured in this way (68). A Richardson plot (69) gave a fractal dimension of approximately 1.29 for step lengths large compared with the particle size but small compared with the dendrite size. Outside these limits the fractal dimension was unity, as expected. Fractal dimensions of deposits are frequently computed in this way, though the method mentioned earlier in this section is more logical.

In experiments on filters, the fractal dimension of a deposit was found (70) to increase steadily with loading time from an initial value of unity, appropriate to a doublet, the smallest "dendrite" for which a fractal dimension could be defined. Values of up to 1.5 were found for deposits formed both by diffusion and impaction, but when loading took place by diffusion this value was reached in about one tenth of the time.

In computer simulations of the formation of two-dimensional agglomerates (71), not attached to fibres, accretion by straight line trajectories, analogous to inertial impaction, gave a fractal dimension of 1.5, whereas formation by diffusion gave a fractal dimension of 1.44.

In experimental work in filters (51), comparisons between the fractal dimension of the boundary of dendrites formed principally by inertial impaction and those formed principally by diffusional deposition, calculated using a Richardson plot, resulted in a fractal dimension of 1.22 being attributed to the former, and 1.37 to the latter.

A study of filter cakes, which are, in a sense, examples of dendrite structures (72), gave fractal dimensions between 1.75 and 1.85. The extreme case studied was a very large agglomerate formed by linear accretion (66), in which particles were given random straight line trajectories, and were assumed to stick on contact. The fractal dimension resulting from the (3-dimensional) simulation was >2.8, which is much bigger than that observed in realistic dendrite simulations, and which corresponds to a deposit of approximately spatially uniform structure.

Effect of loading with liquid aerosols

The theory of capture of liquid aerosols by clean filters is not fundamentally different from that of capture of solid particles; except that in the former case the assumption of spherical particles is usually well-justified. The effect of loading the filter with the aerosol is, however, quite different.

In early work (73, 74) it was observed that captured liquid aerosols were frequently to be found as liquid bridges in the sharp angles between contacting fibres. Experimental results on the behaviour of filters challenged with liquid aerosols could be explained only on the assumption that there were capture sites and storage sites within the filters, and that after loading had taken place some migration of aerosol from the former to the latter occurred. It is likely that the capture sites are the fibre surfaces and that the storage sites are the liquid laden angles (73) illustrated schematically in Fig. 8.16.

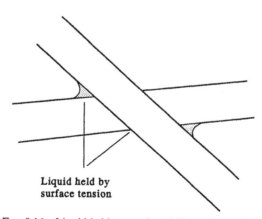

Liquid held by surface tension

FIG. 8.16 Liquid held at a point of fibre–fibre contact.

If this migration process is prevented, and absorption of the droplets by the fibres does not take place, captured liquid drops will form either bead-like structures or structures referred to as "unduloids" (75), illustrated in Fig. 8.17. The degree of spreading will depend upon the values of surface and interfacial energy of the liquid and fibre material

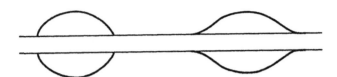

FIG. 8.17   A single fibre holding liquid in the form of a bead (non-zero contact angle) and an "unduloid" (zero contact angle).

but under no circumstances will a uniformly coated fibre occur. Just as a liquid column tends to break up into droplets under the influence of surface tension (76), so will a uniform coating of liquid on a fibre.

A fibre with beads or unduloids on it will have both a higher drag and a higher collection efficiency than a clean fibre; and measurements made of the filtration of liquid droplets using filter fibres with surfaces treated so that either beads or unduloids could form (77), showed that the formation of beads was associated with a higher filtration efficiency.

Experiments carried out on model filters made from parallel wires with a rigid structure (78), heavily loaded with DBP, resulted in an increase in the resistance of the model filters, which held the DBP in the form of unduloids. The resistance of the filter could not be calculated exactly, but of a number of plausible theoretical models fitted to the experimental data, the most successful was one in which the unduloids were represented by short fibres of the same volume, and the entire filter was treated as a polydisperse system.

If the liquid is sufficiently mobile to move to the storage sites, the effect on both efficiency and resistance will be substantially reduced. In addition, a drop of liquid that bridges two fibres that approach each other without touching can reduce its own surface area, and hence its surface energy, by bringing the two fibres closer together (79). This effect requires the filter structure to be non-rigid, and it will be limited in extent by the increased elastic energy of the deformed fibres, the configuration finally adopted by the loaded filter being that of minimum energy. The clumping of fibres resulting from this will reduce both pressure drop and filtration efficiency; and this has been observed (1) with glass fibre and cellulose-glass filters.

Other measurements (80), of the performance of glass fibre filters loaded with submicrometre aerosols of oil, showed a steady increase in pressure drop with load, and a reduction in penetration when the load was sufficient to cause a 20% increase in pressure drop. At higher loads, however, the penetration increased, by a factor of about 2 when the pressure drop was doubled and by a factor of about 4 when it was trebled. This deterioration in performance was attributed to a gross alteration in structure including blocking of parts of the filter by deposited liquid.

Time-dependent effects in filter clogging

All of the work with solid aerosols discussed so far has been based on the rigid dendrite model. However, a real dendrite is not a static structure, and the forces that cause a particle to adhere to a surface also act between a surface and the particles that form part of a dendrite, and between one particle and another within the dendrite. Since forces at a microscopic level are largely attractive, the dendrites will tend to collapse. The collapse of a dendrite will reduce its effect on both the mechanical filtration efficiency and the pressure drop of a filter. It has been suggested that collapse should also cause a slight increase in the fractal dimension of the dendrite (70).

The effect of dust loading on the performance of electrically charged filters

Loading electrically charged filters with dust produces two, sometimes contrary, effects. The dust will tend to produce dendrites just as it does in mechanical types of material; though the electric forces may affect the form (81), development and stability of their structure. One effect of the dendrites will be to increase the filtration efficiency by mechanical means and to increase the airflow resistance, just as they do in mechanical filters (and in filters that act by an externally applied electric field (82)). The second effect of the deposited material, whether in the form of dendrites or not, will be a reduction in filtration efficiency, caused by interaction with the electric charge on the filter, an effect that has no analogy in mechanical filters.

The behaviour observed is complicated by the fact that the two processes will occur together, and the total effect may depend critically on the structure of the particular sample of filter used. Behaviour is often hard to predict and not always repeatable, and frequently the results of loading experiments are difficult to interpret.

An insight into some general aspects of behaviour can be obtained from Fig. 8.18, which shows the behaviour of an electrostatic filter intermittently loaded with aerosol. The effect of loading this filter is to increase the pressure drop and to reduce the penetration; but allowing the filter to rest results in a reduction of the pressure drop, but to a point above its initial value, and an increase in the penetration, also to a point higher than its initial value. The explanation for this behaviour is the formation of dendrites, followed by their collapse, assisted by the electrostatic attraction of the fibre for the particles. In more tightly packed filters both the pressure drop increase and the filtration efficiency increase are better preserved during resting, so that the filter efficiency may never drop below its initial value, and the filter may

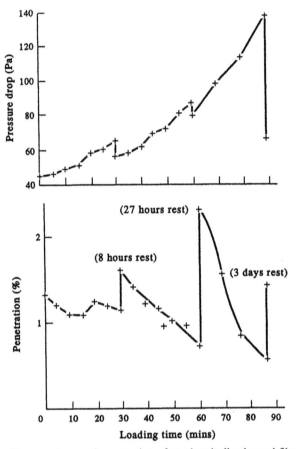

FIG. 8.18   Pressure drop and penetration of an electrically charged filter intermittently loaded with a fine solid aerosol. (© Crown Copyright; by permission of Her Majesty's Stationery Office.)

show signs of marked clogging. In loosely packed filters the charge loss is the dominant process, and the dendrite formation less important; and in these instances, there may be a continuous loss of filtration performance on loading, without any evidence of clogging.

It is very likely that dendrite collapse is a continuous process, which will go on during loading. This means that if loading is carried out quickly, i.e. with a high dust concentration, dendrites will build up, leading to the sort of catastrophic clogging described above. If it is carried out slowly, the dendrites will tend to collapse during loading, and the effects of clogging will proceed at a much slower rate, though clogging will still occur eventually (54). Unfortunately this means that the results of experimental simulations, whereby the effect on the filter of a small dust concentration over a protracted period (likely to occur in practice) is simulated by the effect of a high concentration over a short period (during a laboratory test), must be interpreted with great care.

Empirical theory of charge loss

A significant understanding of the process of charge loss can come about from the application of simple theory to experimental results. The filtration efficiency is assumed to vary exponentially with the filter thickness, which was shown in Chapter 1 to be correct for monodisperse aerosols, and a reasonable approximation if the particle size-variation is not excessive. The effect of aerosol loading in reducing the effectiveness of the electrical charge of the filter is assumed to be independent of where the aerosol is deposited within the filter, which is just a simple linear approximation (83), likely to be valid for small values of penetration. The consequence of these assumptions is that a mass per unit area, M_L, of the loading aerosol alters the penetration of the test aerosol (which may or may not be the same) to a value $P_T(M_L)$, related to its initial value by,

$$\ln\left[\frac{P_T(M_L)}{P_T(o)}\right] = \beta_{TL}M_L. \qquad (8.19)$$

β_{TL} is an index that quantifies the extent to which a particular aerosol causes performance loss of a particular filter material. Where experiments can be carried out accurately, the exponential increase of penetration with M_L, predicted by equation 8.19, is observed.

β-values for a number of different aerosols are shown in Fig. 8.19 (84), which illustrates that the variation in β between different materials loaded with the same aerosol is relatively small, whereas the variation amongst samples of the same material loaded with different aerosols is considerable. The relationship between the β-values for two materials of rather similar geometry is almost linear. This simple relationship can be explained if it is assumed that the aerosol acts by screening the fibre charge. Since the screening charge would be expected to be approximately proportional to the charge being screened, the fractional reduction in single fibre efficiency would be more or less independent of the original charge. If, on the other hand, the process were one of charge neutralisation, the effect would be more marked in the case of filter materials that held a lower level of charge; and this is not observed. In fact measurements of the level of charge on industrial aerosols show that the quantity of aerosol captured does not carry sufficient charge to neutralise electrostatic filters (85).

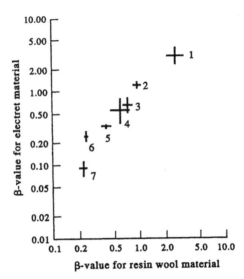

FIG. 8.19 β_{TL} (m^2 g^{-1}), as given by equation 8.19, for a resin wool and an electret filter loaded with industrial aerosols: (1) fine aerosol component of foundry fettling fume; (2) fine aerosol component of foundry burning fume; (3) lead smelting fume; (4) lead battery dust; (5) coke oven fume; (6) silica dust; (7) refractory brick dust. (N.B. β, like layer efficiency and quality factor, is a function of air velocity during filtration (84).) (© Crown Copyright; by permission of Her Majesty's Stationery Office.)

Not all results show the pattern described above, because another process can cause efficiency loss in electrostatic filters: the interaction of solvent-containing aerosols like coal tar fume, with filter materials made from soluble polymers. Insoluble polymers are not susceptible.

Loading with solid aerosols

Laboratory experiments have been carried out (86, 87) in which coarse-fibred and fine-fibred electrically charged materials were loaded with monodisperse aerosols of various sizes, generated from sodium chloride. The pressure drop across the fine-fibred material was increased more by the larger solid particles, 0.1 μm in diameter, than by the smaller, 0.06 μm. Solid laboratory-generated aerosols (86) tend to cause penetration to increase, to reach a maximum, and then to decrease. The maximum was reached earlier with finer fibre materials and with coarser aerosols. In the extreme case the maximum could be assumed to be at a negative load, i.e. there was a monotonic decrease in penetration.

Other authors (88, 89, 90) obtained basically similar results, in loading fine-fibred and coarse-fibred materials with fine (mean diameter 0.04 μm) and coarse (mean diameter 1.5 μm) aerosols. The dominant process in the coarse material was charge loss, since aerosol penetration increased monotonically with load, but the mass of coarse aerosol required to achieve any change was several times that of the fine aerosol achieving the same effect (88). In the case of the fine material, loading with the coarse aerosol caused the penetration to decrease monotonically. Any charge loss was masked by the clogging. The fine aerosol caused an initial penetration increase with a decrease at higher loads due to clogging.

Effect of aerosol charge

Similar materials to the above were loaded with aerosols in different charged states: a completely neutral aerosol, and one that was simply "neutralised", i.e. passed through an ionising neutraliser, which reduces the charge of a highly charged aerosol to a level that can approach the low level appropriate to thermodynamic equilibrium. The neutral aerosol (91) was considerably more clogging to both filter materials than was the aerosol with the low charge; and whereas the latter aerosol caused a practically monotonic decrease in filtration efficiency, the penetration of the neutral aerosol reached a maximum, after which it decreased on further loading. Other authors (92) found a neutralised aerosol to cause a greater reduction of filtration efficiency than a neutral aerosol. Filter behaviour may be highly dependent on experimental conditions.

Loading with liquid aerosols

Liquid aerosols do not form structured deposits within a filter, and so the pattern of behaviour of filters on loading with liquids is quite different from that observed with solids. The effects discussed with reference to mechanical filters may occur, along with the further complication of interaction between the liquid and the material of the fibres. The chemical nature of insulating polymers makes them water-repellant, but it encourages the spread over their surface of oily materials, a category that includes relatively few industrial aerosols but several test aerosols: paraffin oil, di-octyl phthalate (DOP) and di-octyl sebacate (DOS). Spreading of the oil screens out the electric charge, but although the resulting reduction in electrostatic energy encourages spreading, the effect is secondary. The spreading of the oil depends principally on interfacial tensions (93), and a complete description would require knowledge of the values of these tensions.

The pattern followed by the aerosol penetration through filters loaded with oily aerosols is not unlike that given by equation 8.19, except that the oil may take a finite time to reach its equilibrium surface coverage.

References

1. SMISSEN, C. E. VAN DER. Loading capacity of aerosol filters for respirators. *Staub Reinhalt. Luft.*, 1971, **31**, 1–5 (in English).
2. NIELSEN, K. A. and HILL, J. C. Particle chain formation in aerosol filtration with electric forces. *AIChE J.*, 1980 **26**, 678–680.
3. BERGMAN, W. *World Filtration Congress 2 Proceedings*, 1979.
4. OAK, M. J. and SAVILLE, D. A. The build up of dendrite structures on fibres in the presence of strong electric fields. *J. Coll. Int. Sci.*, 1980, **76**(1), 259–262.
5. OAK, M. J., SAVILLE, D. A. and LAMB, G. E. R. Particle capture on fibres in strong electric fields. 1. Experimental studies of filter charge fibre configuration and dendritic structure. *J. Coll. Int. Sci.*, 1985, **106**(2), 490–501.
6. KANAOKA, C., EMI, H., HIRAGI, S. and MYOJO, T. Morphology of particulate agglomerates on a cylindrical fibre and collection efficiency of a dust-loaded filter. 1986, *Aerosols, Formation and Reactivity, 2nd Int. Conf. Berlin*, 674–677, Pergamon, 1986.
7. KUWABARA, S. The forces experienced by a lattice of elliptic cylinders in a uniform flow at small Reynolds numbers. *J. Phys. Soc. Japan*, 1959, **14**(4), 522–527.
8. JUDA, J., CHROSCIEL, S. and NOWICKI, M. The influence of some filter parameters on the pressure loss increase in the filtration process. *Staub Reinhalt. Luft.*, 1973, **33**, 159–162.
9. SHUCOSKI, A. C., GERACI, V. A., TURNER, M. C. and CAMERON, J. E. Characterization of absolute-rated filter material for oil and gas production filtration. *Advances in Filtration and Separation Technology*, Vol. 1, American Filtration Society, Oct. 30–Nov. 1, 1989, Houston, Texas.
10. DULLIEN, F. A. L. Maximizing the capacity and life of depth-type filters. *Can. J. Chem. Eng.*, 1989, **67**, 689–692.
11. BROWN, R. C. A simple macroscopic model of the clogging of depth filters, to be published.
12. HERZIG, J. P., LECLERC, D. M. and LEGOFF, D. Flow of suspensions through porous media. Application to deep bed filtration. *Ind. Eng. Chem.*, 1970, **62**(5), 8–35.
13. DVUKHIMENNYI, V. A., KIRSCH, A. A., STECHKINA, I. B. and USHAKOV, E. N. Efficiency variation and pressure drop of aerosol filters during accumulation of a sediment of solid particles on fibres. *Theor. Found. Chem. Eng.*, 1985, **19**(5), 428–433.
14. BATCHELOR, G. K. *Introduction to Fluid Dynamics*, CUP, Cambridge, 1967.
15. MILLIKAN, R. Coefficient of slip in gases and the law of reflection of molecules from the surfaces of gases and liquids. *Phys. Rev.*, 1923, **21**(3), 217–238.
16. DAVIS, A. M. J., O'NEILL, M. E., DORREPAAL, J. M. and RANGER, R. B. Separation from the surfaces of two equal spheres in Stokes flow. *J. Fluid Mech.*, 1976, **77**(4), 625–644.
17. HORVATH, H. The sedimentation behaviour of non-spherical particles. *Staub Reinhalt. Luft.*, 1974, **34**, 197–202 (in English).
18. DAHNEKE, B. Viscous resistance of straight chain aggregates of uniform spheres. *Aerosol Sci. Technol.*, 1982, **1**, 179–185.
19. LEE, C. T. and LEITH, D. Drag force on agglomerated spheres in creeping flow. *J. Aerosol Sci.*, 1989, **20**, 503–573.
20. DAVIES, C. N. Particle fluid interaction. *J. Aerosol Sci.*, 1979, **10**, 477–513.
21. CHENG, Y. S., ALLEN, M. D., GALLEGOS, D. P., YEH, H. C. and PETERSON, K. Drag force and slip correction of aggregate aerosols. *Aerosol Sci. Technol.*, 1988, **8**, 199–214.
22. PENDSE, H., TIEN, C. and TURIAN, R. M. Drag force measurement of single spherical collectors with deposited particles. *AIChE J.*, 1981, **27**, 364–372.
23. HAPPEL, J. and BRENNER, H. *Low Reynolds Number Hydrodynamics*, 4th Edition, Nijhoff, Dordrecht, 1986.
24. O'NEILL, M. E. A sphere in contact with a plane wall in slow linear shear flow. *Chem. Eng. Sci.*, 1986, **23**, 1293–1298.
25. GOLDMAN, A. J., COX, R. J. and BRENNER, H. Slow viscous motion of a sphere parallel to a plane wall II—Couette flow. *Chem. Eng. Sci.*, 1967, **22**, 653–660.

26. MEDJIMOREC, V., OKUYAMA, K. and PAYATAKES, A. C. Estimation of drag forces acting on particle dendrites. *J. Coll. Int. Sci.*, 1981, **82**, 543–559.

27. RAMARAO, B. V. and TIEN, C. Calculations of drag forces acting on dendrites. *Aerosol Sci. Technol.*, 1988, **8**, 81–95.

28. RADUSHKEVICH, L. V. Kinetics of the formation and growth of aggregates on a solid obstacle from a flow of colloid particles. *Koll. Zh.*, 1964, **26**, 235–240 (in Russian), H.S.E. Translation 8210, 1979.

29. PAYATAKES, A. C. and TIEN, C. Particle deposition in fibrous media with dendrite-like pattern: a preliminary model. *J. Aerosol Sci.*, 1976, **7**, 85–100.

30. PAYATAKES, A. C. Dendrites and air filter efficiency. *Proc. Filtr. Soc.*, Nov. 1976, 602–607.

31. PAYATAKES, A. C. Model of transient aerosol particle deposition in fibrous media with dendritic pattern. *AChE J.*, 1977, **23**, 192–202.

32. PAYATAKES, A. C. and GRADON, L. Dendritic deposition of aerosols by convective Brownian motion for small, intermediate and high Knudsen numbers. *AChE J.*, 1980, **26**(3), 443–454.

33. TIEN, C., WANG, C. S. and BAROT, D. T. Chainlike formation of particle deposits in fluid–particle separation. *Science*, 1977, **196**, 983–985.

34. WANG, C. S., BEIZAIE, M. and TIEN, C. Deposition of solid particles on a collector; formulation of a new theory. *AIChE J.*, 1977, **23**, 879–889.

35. BEIZAIE, M., WANG, C. S. and TIEN, C. A simulation model of particle deposition on single collectors. *Chem. Eng. Comm.*, 1981, **13**, 153–180.

36. KANAOKA, C., EMI, H. and MYOJO, T. Simulation of the growing process of a particle dendrite and evaluation of a single fibre collection efficiency with dust load. *J. Aerosol. Sci.*, 1980, **11**, 77–389.

37. KANAOKA, C., EMI, H. and TANTHAPANICHAKOON, W. Convective diffusional deposition and collection efficiency of aerosol on a dust-loaded fiber. *AChE J.*, 1983, **29**(6), 895–902.

38. BAHNES, T. and SCHOLLMEYER, E. Computer simulation of the filtration processes in a fibrous filter collecting polydisperse dust. *J. Aerosol Sci.*, 1987, **17**(2), 191–200.

39. BAUMGARTNER, H. and LOEFFLER, F. Three-dimensional numerical simulation of the deposition of polydisperse aerosol particles on filter fibres—extended concept and preliminary results. *J. Aerosol. Sci.*, 1987, **18**, 885–888.

40. CAI, J. Instationary filtration due to dendrites. *Proceedings of the 3rd International Aerosol Conference, Kyoto, Sept. 24–27 1990*, 715–719, Pergamon, 1990.

41. BAROT, D. T., TIEN, C. and WANG, C. S. Accumulation of solid particles on single fibers exposed to aerosol flows. *AIChE J.*, 1980, **26**(2), 289–292.

42. BHUTRA, S. and PAYATAKES, A. C. Experimental investigation of dendritic deposition of aerosol particles. *J. Aerosol Sci.*, 1979, **10**, 445–464.

43. OKUYAMA, K. and PAYATAKES, A. C. Comparison between theory and experiment in dendritic deposition; a revision. *J. Aerosol Sci.*, 1981, **12**, 269–274.

44. DAHNEKE, B. and PADILYA, D. Comments on paper on experimental deposition of aerosol particles. *J. Aerosol Sci.*, 1980, **11**, 567–569.

45. MOHRMANN, H. and MARCHLEWITZ, W. Loading of fiber filters with aerosols. *Staub Reinhalt. Luft.* 1975, **34**, 91–93 (in English).

46. BILLINGS, C. E. Effects of particle accumulation on aerosol filter life. *Proceedings of the 9th AEC Air Cleaning Conference*, 656, 1966.

47. DAVIES, C. N. The clogging of fibrous aerosol filters. *Aerosol Science*, 1970, **1**, 35–39.

48. MYOJO, T., KANAOKA, C. and EMI, H. Experimental observation of collection efficiency of a dust-loaded fibre. *J. Aerosol Sci.*, 1984, **25**, 483–489.

49. KANAOKA, C. and HIRAJI, S. Pressure drop of air filter with dust load. *J. Aerosol Sci.*, 1990, **21**, 127–137.

50. EMI, H., WANG, C. S. and TIEN, C. Transient behaviour of aerosol filtration in model filters. *AIChE J.*, 1982, **28**, 397–405.

51. TROTTIER, R. A., STENHOUSE, J. I. T. and KAYE, B. H. Possible link between the fractal structure of dust capture deposits in a fibrous filter and loading effects. *5th Aerosol Society Conference Proceedings*, 71–76, 1991.

52. STENHOUSE, J. I. T. and TROTTIER, R. A. The loading of fibrous filters with submicron particles. *J. Aerosol Sci.*, 1991, **22**, S777–S780.

53. LETOURNEAU, P., MULCEY, P. and VENDEL, J. Effect of dust loading on the pressure drop and efficiency of HEPA filters. *Filtr. Sep.*, July 1987, 265–267.

54. FIRST, M. W. and RUDNICK, S. N. Performance of 1000 and 1800 cfm HEPA filters on long exposure to atmospheric dust loads. *2nd World Filtration Congress Proceedings*, 283–289, 1979.

55. YOSHIOKA, N., EMI, H., YASUNAMI, I. M. and SATO, H. Filtration of aerosols through fibrous packed bed with dust loading. *Chem. Eng. Tokyo*, 1969, **33**, 1013–1018.

56. LEIBOLD, H. and WILHELM, J. G. Investigations into the penetration and pressure drop of HEPA filter media during loading with submicron particles at high concentrations. *J. Aerosol Sci.*, 1991, **22**, S773–S776.

57. NOVICK, V. J., MONSON, P. R. and ELLISON, P. E. The effect of solid particle mass loading on the pressure drop of HEPA filters. *J. Aerosol Sci.*, 1992, **23**(6), 657–665.

58. STAFFORD, R. G. and ETTINGER, H. J. Filter efficiency as a function of particle size and air velocity. *Atmos. Environ.*, 1972, **6**, 353–362.

59. TSIANG, R. C., WANG, C. S. and TIEN, C. Dynamics of particle deposition in model filters. *Chem. Eng. Sci.*, 1982, **37**, 1661–1673.

60. MIYAGI, T. Viscous flow at low Reynolds number past an infinite row of equal circular cylinders. *J. Phys. Soc. Japan*, 1958, **13**, 493–496.

61. RAMARO, B. V. and TIEN, C. Stochastic simulation of aerosol deposition in model filters. *AIChE J.*, 1988, **34**, 253–262.

62. DAHNEKE, B. The capture of aerosol particles by surfaces. *J. Coll. Int. Sci.*, 1971, **37**(2), 242–353.

63. STENHOUSE, J. I. T., BROOM, G. P. and CHARD, N. J. T. Dust loading characteristics of high inertial fibrous filters. *Am. Ind. Hyg. Assoc. J.*, 1978, **39**, 219–225.

64. MANDELBROT, B. B. *Fractals, Form Chance and Dimensions*, Freeman, San Francisco, 1977.

65. CRILLY, A. J., EARNSHAW, R. A. and JONES, K. *Fractals and Chaos*, Springer-Verlag, New York, 1991.

66. MEAKIN, P. Accretion processes with linear particle trajectories. *J. Coll. Int. Sci.*, 1985, **105**, 240–246.

67. ROGAK, S. N. and FLAGAN, R. C. Stokes drag on self-similar clusters of spheres. *J. Coll. Int. Sci.*, 1990 **134**(1), 206–218.

68. ENSOR, D. and MULLINS, M. E. The fractal nature of dendrites formed by the collection of particles on fibres. *Part. Charact.* 1985, **2**, 77–78.

69. KAYE, B. H. *A Random Walk Through Fractal Dimensions*, VCH, Weinheim, 1989.

70. KANAOKA, C., HIRAGI, S. and TANTHAPANICHAKOON, W. Fractal analysis of fluid drag acting on a fibre with dust load. *Proc. 3rd Int. Aerosol Conf., Kyoto, 1990 Sept. 27–27*, Pergamon, 1990.

71. MEAKIN, F. Structural readjustment effects in cluster aggregation. *J. Phys.*, 1985, **46**, 1543–1552.

72. BAYLES, G. A., KLINZING, G. E. and CHIANG, S. H. Fractal mathematics applied to flow in porous systems. *Part. Syst. Charact.*, 1989, **6**, 168–175.

73. LANGMUIR, I. and BLODGETT, K. B. (see LAMER, V. K. and DROZIN, V. G.). Filtration of monodisperse solid aerosols. *Proceedings of the Second International Conference on Surface Activity*, **3**, 49–55, 1957.

74. MOHRMANN, H. Loading of fibrous filters with aerosols consisting of liquid particles. *Staub Reinhalt. Luft.*, 1970, **30**(8), 1–6 (in English).

75. ROE, R. J. Wetting of fine wires and fibres by a liquid film. *J. Coll. Int. Sci.*, 1975, **50**(1), 70–79.

76. RAYLEIGH. On the instability of a cylinder of viscous liquid under capillary force. *Phil. Mag.*, 1892, **34**, 145–154.

77. FAIRS, G. L. High efficiency fibre filters for treatment of fine mists. *Trans. Inst. Chem. Eng.*, 1958, **36**, 476–485.

78. KIRSCH, A. A. Increase in pressure drop in a model filter during mist filtration. *J. Coll. Int. Sci.*, 1978, **64**, 120–125.

79. ERIKSSON, J. C., LJUNGGREN, S. and ODBERG, C. Adhesion forces between fibres due to the capillary condensation of water vapour. *J. Coll. Int. Sci.*, 1992, **152**(2), 368–375.

80. PAYET, S., BOULAUD, P., MADELEINE, G. and RENOUX, A. Dynamic filtration of liquid aerosols. *Proceedings of the 5th World Filtration Congress*, 617–623, 1990.

81. HENRY, F. S. and ARIMAN, T. Numerical modelling of electrically enhanced fibrous filters. In *Fluid Filtration; Gas* (ed. R. R. RABER), Volume 1, American Society for Testing and Materials, Philadelphia, 1986.

82. GRADON, L. Influence of electrostatic interations and slip effect on aerosol filtration efficiency in fibre filters. *Ind. Eng. Chem. Res.*, 1987, **26**, 306–311.

83. KANAOKA, C., EMI, H. and ISHIGURO, T. Time dependency of collection performance of electret filters. *1st AAAR Conference, 1984* (ed. B. Y. H. LIU et al.), 614–616, Elsevier, NY, 1984.

84. BROWN, R. C., WAKE, D., GRAY, R., BLACKFORD, D. B. and BOSTOCK, G. J. Effect of industrial aerosols on the performance of electrically charged filter material. *Ann. Occ. Hyg.*, 1988, **32**, 271–294.

85. JOHNSTON, A. M., VINCENT, J. H. and JONES, A. D. Measurement of electric charge for workplace aerosols. *Ann. Occ. Hyg.*, 1985, **29**(2), 271–284.

86. LATHRACHE, R., FISSAN, H. and NEUMANN, S. Dynamic behaviour of electrostatically charged fibrous filters. *Aerosols, Formation and Reactivity, 2nd Int. Aerosol Conf., Berlin, 1986*, 712–715, 1986.

87. LATHRACHE, R. and FISSAN, H. Grundelegende untersuchungen zum abscheideverhalten der electretfilter Teil 3: Anderung der filtrationseigenschaften durch partikelbeladung. *Staub Reinhalt. Luft.*, 1989, **49**, 407–411.

88. BAUMGARTNER, H. P. and LOEFFLER, F. The collection performance of electret filters in the particle size range 10 nm–10 μm. *J. Aerosol Sci.*, 1986, **17**(3), 438–445.

89. JODEIT, H. and LOEFFLER, F. The influence of electrostatic forces on particle collection in fibrous filters. *J. Aerosol Sci.*, 1984, **15**, 311–317.

90. SHIMOKOBE, I. and INOUE, M. Collection performance of electret filter for particles. *J. Aerosol Res., Japan*, 1986, **1**(3), 186–192.

91. BAUMGARTNER, H. and LOEFFLER, F. Experimental and theoretical investigation of the time-dependent collection performance of electret filters. In *Aerosols, Performance and Reactivity, 2nd Int. Aerosol Conf., Berlin, 1986*, 708–771, 1986.

92. SCHURMANN, G. and FISSAN, H. J. Fractional efficiency of an electrostatically spun polymer fibre filter. *J. Aerosol Sci.*, 1984, **15**, 3127–230.

93. DEFAY, R. *Surface Tension and Adsorption*, Longmans, London, 1966.

CHAPTER 9

Filter Testing

Introduction

Commercially produced filters are frequently subjected to quality control tests by the manufacturer. In addition a user may apply a special test to a filter intended for a specific use. Most of the data presented in the preceding chapters are the results of tests carried out with the purpose of increasing general understanding of filter behaviour. Tests may deal with a variety of properties such as size and weight, physical robustness, or resistance to attack by chemical agents, but the tests to be treated in this chapter are solely those that refer to a filter's performance as such.

Pressure drop measurement

The quantification of filter performance involves measurement of both the pressure drop across the filter and the penetration of aerosol through it. The former test is relatively easy to perform; and results can be generalised to predict behaviour at any filtration velocity using Darcy's law, with correction terms if necessary. A complication is that the filter housing can introduce an inertial component into the flow, with the result that, although the filter may obey Darcy's law, the pressure drop across the entire unit of filter plus housing may be higher than Darcy's law would predict (1).

If filters are to be used in non-standard conditions of pressure or temperature, the effect of these variables on pressure drop measurement may become important; in particular, the effects of aerodynamic slip may become apparent at low pressures if the fibre size is sufficiently small. The viscosity of the air varies as the square root of the absolute temperature and so pressure drop will increase slightly with temperature. Viscosity is independent of pressure, but at high pressures the difference between volume flow rate, on which the simple filtration equations are based, and mass flow rate, which is the parameter measured by certain types of flowmeter, becomes significant (1).

Area weight measurement

The measurement of filtration efficiency involves many more problems, since it depends on air velocity, particle size and, in some cases, particle charge, though in certain instances a measure of area weight or filter thickness may give a sufficient indication of performance for a manufacturer's quality control purposes. A popular method of monitoring the area weight of paper filters is to scan the filter with a β-particle emitter and detector (2), the degree of attenuation of the radiation being related to the weight of material between the two. Incorporation of this principle into an automated device is straightforward.

General problems of aerosol penetration measurement

A test may simply be required to indicate whether or not a filter will give a performance at least equal to some previously set standard, or the measurement may be aimed at getting a complete picture of aerosol penetration as a function of particle size. The approach taken in this chapter will be to cover the principles of filter testing rather than to give a catalogue of all existing test methods, though descriptions of and comparisons among test methods, past and present, are given elsewhere (3, 4, 5).

Sample uniformity and consistency

Testing filters requires a sample of filtered aerosol to be compared with one un-filtered; and there are two basic methods of obtaining the samples. An aerosol may be continually passed through the filter, whilst samples are taken upstream and downstream, as illustrated in Fig. 9.1; or the aerosol may be passed through the filter intermittently, and by-passed for comparison, as shown in Fig. 9.2. The former method is particularly suitable for *in-situ* filter tests, the latter for laboratory tests.

Each method has its own particular problems, but a requirement of both is that the nature and concentration of the test aerosol should not vary with time, unless, in the former situation, two detectors or samplers with identical characteristics are used simultaneously, or, in the latter, a rapid-switching procedure is employed (6).

In the two sampler method it is ideal if the sampling probes are isokinetic (7), but if they have imperfect but identical characteristics these should have no bearing on the results. Unless the filter is uniform, however, it is essential that the aerosol be properly mixed both upstream and downstream of the filter. In the case of a leaking or

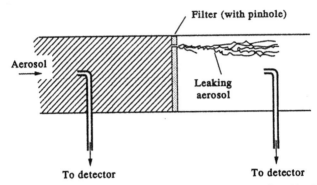

FIG. 9.1   Upstream–downstream sampling method for filter testing. The figure also shows a pinhole and the localised high concentration of leaking aerosol downstream.

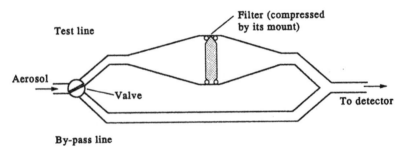

FIG. 9.2   By-pass and switching method for filter testing. The figure also shows the filter being compressed by a baldy designed filter-mount.

perforated filter the unmixed aerosol concentration immediately downstream of the filter will be far from homogeneous, as shown in Fig. 9.1; and the average may be seriously over-estimated or under-estimated. Mixing can be aided by the turbulence resulting from a restricted orifice or elbow (8), by passing the aerosol through a system of ducting that ensures "cut and mix", or by using an auxiliary fan to disturb the airflow.

Filter scanning

An alternative strategy is to prevent downstream mixing so that the test can be used to localise leaks (9). In this method a sampling probe is passed in a scanning motion close to the surface of the filter so that air passed through different parts of the filter does not mix before it is sampled. The probe must be small enough to allow the leak to be located precisely (10), and the entry velocity to the probe should be neither lower than the filtration velocity nor so much higher that it samples air emanating from a wide area of filter.

Line losses

In the switching test method it is essential that the by-pass line has the same extent of particle loss as that caused by the filtering line, apart from the action of the filter itself; and the best way of achieving this is to minimise line losses. In both methods the most likely losses are those caused by aerodynamic effects, either inertial impaction or gravitational settling, and so there are rather more problems with the use of coarse aerosols. Testing with these is often better carried out in conditions of downflow, to eliminate, as far as possible, particle losses by settling (4, 11).

Uniformity of air velocity through the filter

In each case it is necessary that the air should pass through the filter at the same velocity everywhere, since filtration is a velocity-dependent process. Many filters, particularly high-resistance filters, are good flow-straighteners, ensuring that the air flows uniformly through them; but in the situation where aerosol passes from a narrow duct into a much wider filter test chamber in order to pass through a filter of relatively low resistance, it is necessary that the airflow is not required to diverge too abruptly. A badly designed test system may result in the aerosol being channelled through the central region of the upstream parts of the filter, before the filter is able to exert its flow-straightening properties. The effect of this is shown in Fig. 9.3, which is taken from measurements made using 5 μm aerosol through a coarse porous foam filter tested in layers (12). Channelling has caused the first layer of the filter to remove aerosol over a more limited area and therefore at a higher velocity that the rest, and so the standard graph of penetration (logarithmically) against thickness (linearly) does not pass through the origin, even though the test aerosol is monodisperse.

It is essential that the filter be properly sealed in its mount, so that leakage does not occur. However, a badly designed mount may obscure part of the filter, with the result that the test aerosol passes through the remainder of the filter at a higher velocity than intended, affecting both pressure drop and filtration efficiency. A bad mount may, as sketched in Fig. 9.2, compress the filter at its edges; and this will also alter the filter's performance.

Re-circulating filters

A method that can be applied to re-circulating air-cleaning filters is to sample the air within the (closed) contaminated region during filtration. A steady drop in concentration with time should be observed which, in the case of a monodisperse aerosol completely mixed after

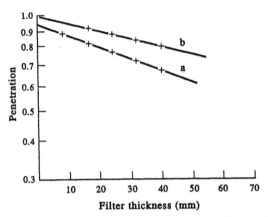

FIG. 9.3 Effect of non-uniform flow in a filter on the observed penetration curve (a) non-uniform flow, (b) uniform flow (12). (© Crown Copyright; by permission of Her Majesty's Stationery Office.)

filtration, should be an exponential decay (13). If the aerosol is polydisperse the pattern of change of concentration with time will be similar to that of concentration change with depth followed by conventional filters tested with polydisperse aerosols. The results obtained lack the rigour of other methods, being dependent on the sampling site and filter position, but the method can be used for comparing the performance of materials (14).

Aerosol detectors

The aerosol assessment method may involve sampling followed by analysis, in which case general aerosol assessment techniques or methods tailored to the particular test aerosol may be used. Membrane filter samples may be analysed gravimetrically, by microscopy or by Coulter analysis. Cascade impactors, which size particles according to their aerodynamic diameter, can be used, as can diffusion batteries (15) which size according to diffusivity. Fluorescent or coloured aerosol can be analysed on membrane filter samples using fluorimetry or colorimetry.

In addition to these, real time methods may be incorporated into filter test systems (16). Many real time instruments depend on optical scattering, for example optical counters respond to the scattered light caused by single aerosol particles passing through the counter's sensing volume; and the height of the pulse can, to some extent, be related to the size of the particle.

Optical particle counters must be distinguished from optical photometers, which measure the intensity of scattered light coming from an entire aerosol (17); and flame photometers, which are used to quantify the sodium chloride aerosols used in a standard test (18), the sodium D-line being particularly bright.

The aerodynamic behaviour of an aerosol can be quantified from measurements of the time of flight of particles crossing two light beams, in an accelerating airflow (19), though in the instrument described the property measured is not exactly the aerodynamic diameter because the conditions are non-Stokesian. There are also electrostatic classifiers, which measure the electrical mobility of singly charged particles, and which are limited to the assessment of submicrometre particles. The detection of fine particles is often aided by condensation nucleus counters, though these are not strictly aerosol detectors since they must be used with an ancillary quantification system.

Testing with monodisperse aerosols

The specific advantage of carrying out tests with monodisperse aerosols is that the layer efficiency can be obtained from a single measurement,

$$P = \exp(-\alpha h) \tag{9.1}$$

where P is the experimentally measured penetration, h is the filter thickness and α is the layer efficiency. Since the thickness of homogeneous material is proportional to the pressure drop, a similar functional relationship holds, with Δp replacing h, and with the quality factor, QF, replacing α,

$$P = \exp(-QF \, \Delta p). \tag{9.2}$$

Both the layer efficiency and the quality factor are independent of the thickness of the filter material tested, but both vary with the filtration velocity. However, even if the velocity is fixed, the results of different tests using aerosols of the same geometric size may not always be the same. If a filter acts purely by interception the capture efficiency is specified by the interception parameter (equation 4.5) and therefore by the geometric size, and so the penetrations of all aerosols with the same geometric size will be the same; but if a filter that acts by inertial impaction is tested with two aerosols of the same physical size but of different densities, the results will differ, since the efficiency of the filter depends not on geometric size but on aerodynamic diameter.

Methods of production of monodisperse aerosols

Aerosol produced from liquid droplets

Monodisperse aerosols may be generated in a variety of ways (20), one of which starts with the production of liquid droplets of a constant size, so that if the liquid is a solution of the material required for the aerosol, the droplets can be made to evaporate, leaving the monodisperse aerosol as a solid residue. The size of the particles can be adjusted by altering the concentration of the solution.

This aerosol generation method can be put into practice by means of a spinning top generator (21) in which liquid is applied at a steady rate to the centre of a rotating flat top, to appear as droplets in the form of breaking filaments regularly spaced around the circumference.

A second generator of this type is the vibrating orifice generator (22). In this device a fine filament of liquid extruded through a small orifice at a constant rate is made to break up into identically sized droplets by the application of a mechanical vibration produced by a piezo-electric crystal acting under the influence of a r.f. electric field. The droplets are then dispersed and diluted to prevent coagulation.

The primary droplets produced by either device are of the order of tens of micrometres in diameter. Since solvents are invariably contaminated by small quantities of soluble residue, there is a practical limit to the degree of dilution that can be effected, with the result that the diameter of monodisperse particles produced has a lower limit of about 1 μm.

The droplets carry an electric charge appropriate to particles of their size, and as the solvent evaporates this charge becomes confined to a much smaller droplet, where it may be extremely troublesome, even causing the Rayleigh limit (23, 24) to be exceeded (which occurs when the repulsive force caused by the electric charge, tending to disrupt the droplet, overcomes the force of surface tension holding it together). Even if this does not occur, the charge may have a large effect on the electrical behaviour of the particle produced. A neutraliser is essential when small particles are generated in this way, and more will be said about this sort of device below. A second possible problem is that if evaporation is too rapid, the particles may take the form of cenospheres (25), with properties rather different from those of dense spheres.

Aerosols produced from condensation nuclei

A further method is the production of condensation aerosols by the Sinclair-LaMer type of generator. The principle of operation of this device is (26) that an aerosol consisting essentially of clean air apart from a high concentration of condensation nuclei is mixed with a highly concentrated vapour. As condensation takes place, an aerosol with a very narrow range of sizes is produced. Particles may range in size from about 0.01 to > 1 μm; but a limitation is that only particles of relatively volatile materials such as oils and waxes can be produced as aerosols in this way. Since the mechanism of production involves the accretion of material around a small nucleus, these aerosols tend to carry a low charge, characteristic of the nucleus around which they condensed.

Polystyrene latex aerosols

Monodisperse particles are commercially available in the form of polystyrene latex, which can be generated as an aerosol by the use of a pneumatic atomiser. A latex particle is contained within a droplet of spraying liquid, and as the latter evaporates the particle is left airborne. The particles will tend to be distributed randomly within the liquid and so if the atomised spray is monodisperse, the probability of finding n particles within a droplet will be given by the Poisson distribution,

$$P_s(n) = \frac{\exp(-\mu_s)\mu_s^n}{n!} \tag{9.3}$$

where, μ_s, the mean number of particles, is the product of the droplet volume and the concentration of latex particles in the generating solution. If μ_s is high there will be a large number of doublets and triplets, and if it is low there will be a large number of empty droplets. Pure liquid droplets would evaporate to nothing, but normally a stabiliser is present, which keeps the latex particles in suspension. However, even if this is removed (27) a very small residual particle is formed from the evaporation of those particles that do not contain latex spheres. A large concentration of these can give a spurious signal in an optical counter even if each, singly, is too small to scatter sufficient light to do so.

If, as is usually the case, the particles are generated from polydisperse spray, the number distribution will have a spread that depends on the form of the size distribution of the spray, and which increases with the spread of size of the droplets of the generating liquid (28).

Aerosols produced by classification

Monodisperse aerosols can be produced by extracting from a poly-disperse aerosol particles of one size only. In principle any method of selection could be used for this purpose, but the most popular is that of electrostatic classification (29, 30). In this method, which is limited in application to submicrometre aerosols, a primary aerosol is produced, often by atomisation, and the particles are then exposed to ions. Very small particles rarely carry more than a single charge, and so the charged particles may be selected on the basis of their electrical mobility, using the device indicated schematically in Fig. 9.4. If a fine mobility-fraction is selected, these will be almost monodisperse and singly charged. The absolute level of charge of these aerosols is as low as a finite charge can be, but the small size of the particles means that their electrical mobility is extremely high. If this is troublesome an aerosol neutraliser may be used.

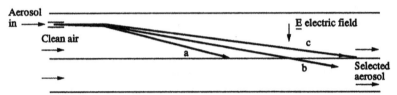

FIG. 9.4 Schematic diagram of an electrostatic classifier. The figure shows classification by a device with planar geometry. The standard device uses cylindrical geometry, but the principle of action is the same: (a) trajectory of rejected particle with high electrical mobility; (b) trajectory of selected particle with required electrical mobility; (c) trajectory of rejected particle with low electrical mobility.

Fibrous aerosols

The production of monodisperse particles of non-spherical, in particular fibrous, shape, has not progressed as far as the production of spheres. It is a much more difficult technical problem because two parameters, length and diameter, need to be regulated. Fibres can be spun from glass or polymer, by a process in which the fibre diameter can be controlled, though the very long fibres spun in this way must be cut to specified lengths in order to be suitable for aerosol generation.

The fibres may be prepared in approximately the required length by milling, followed by classification according to length and diameter. The aerodynamic diameter of a fibre depends much more on its diameter than on its length, and so diameter-classification can be carried out by settling or centrifugation, the recommended method (31) being repeated gravitational settling in water specially treated to remove any flocculating agents. Selection according to length can be carried out by sieving, though the angle of inclination of the fibres is critical. Alternatively the use of angle of descent as a selection method has been proposed (32).

Yarns or wool of long glass fibres with diameters as small as 2.5 and 0.5 μm respectively can be made into short fibres by mechanical alignment and potting in histological medium, which is then sliced with a microtome (31, 33) to give fibres of 20–50 μm in length. The potting medium is removed by ashing and the residue by elutriation.

Monodisperse carbon fibres can be produced by chopping long aligned fibres with a laser (34), which results in a fairly tight length distribution, but the rate of production is low and the method appears to be limited to fibres of 10 μm or more in diameter.

The electrostatic classification process described above has been used in the separation of fibres from isometric particles, and in the classification of fibres according to their aspect ratio (35, 36, 37). A simplified calculation based on diffusion charging of particles predicts that a fibre with an aspect ratio of 100 will have an electrical mobility greater by a factor of 1.8 than that of a spherical particle with the same aerodynamic diameter.

The condensation nucleus type of generator can be used to produce monodisperse crystals of non-spherical shape if the generating material sublimes. (Condensation to a liquid and subsequent solidification invariably produces isometric particles.) The shape of the crystals is fixed by the nature of the generating material, and monodisperse fibres are produced if caffeine is used as the aerosol-generating material (38).

The process used to etch microcircuit components and to produce model filters (see Chapter 2) can be applied to the production of monodisperse particles in a wide range of shapes, by the appropriate choice of template (39). Fibres are amongst the simplest and most useful that could be produced in this way.

Quantification and minimisation of random errors

If the particle size of an aerosol is prescribed, the detection method need not be highly size selective, and a simple counting technique is quite adequate. Since aerosol penetration is calculated as the quotient of two counts, that in the filtered line and that in the unfiltered, particle statistics can be used both to quantify the error and to suggest a counting technique that will minimise it.

If particles are produced in constant airflow at a constant rate, as they are when made by a vibrating orifice type of aerosol generator, a random sample of air selected for particle counting will have a likely count distribution given by Poisson statistics, in which the mean and the variance are equal. This means that the coefficient of variation, the quotient of the standard deviation and the mean, will be equal to the square root of the average count, N_t in unit time. The same reasoning can be applied to the filtered aerosol, and the likely error in an estimate of the quotient of the two can be given by combining the two errors as a root sum of squares. If the time devoted to a measurement, t_o, is a constant, the optimal division of time between the counting of the filtered and unfiltered aerosols is in the ratio of the reciprocal of the square root of the penetration, P, and the minimum error is (12),

$$\text{Minimum Likely Error} = \frac{1 + P^{1/2}}{(N_t t_o P)^{1/2}}. \qquad (9.4)$$

Rigorous application of this technique requires the answer to be known in advance, but a time division based on a reasonable estimate is an improvement on equal division. In the case of very low penetrations equal division of time between the two aerosols results in a likely error that is greater than the minimum possible by a factor of up to $\sqrt{2}$.

This philosophy has been used in the testing of HEPA filters (40), during which tests the downstream aerosol was sampled for six times as long as the upstream. The variation of minimum error with the number of particles counted has led some workers (41) to advocate the counting of small penetrating particles (0.1 μm say), in order to validate a filter, rather than the larger (0.5 μm say) and less frequent particles on which the standard of performance is based, because correlation is good and the higher count substantially reduces random error.

Application to layer efficiency or quality factor

Sometimes the layer efficiency or quality factor of a filter material is required rather than the performance of a particular filtering device; and in this case it is useful to know what thickness of material should be used for the most accurate result. If a very thin filter is used, filtered and un-filtered aerosols will be so similar in counts that random variations may lead to an error even in the sign of the measured layer efficiency. If, on the other hand, the sample is extremely thick, so few penetrating particles will be counted that statistical errors in the estimation of penetrating aerosol will dominate. Application of the same philosophy as above to this problem leads (42) to the prediction that the ideal thickness is that which gives rise to a penetration of about 7.7%. Errors are close to the minimum values, provided that the sample gives a penetration between about 2 and 20%, but outside these limits errors increase considerably.

Automated methods of obtaining penetration curves with monodisperse aerosols

A problem with monodisperse aerosol testing is that a test gives just a single experimental point, whereas often what is required is a curve of the penetration of aerosol through the filter as a function of particle size. An automated system can, therefore, be of great value. The electrostatic classifier method for submicrometre particles is the easiest device to automate because the aerosol reaching the classifier itself has a range of particle sizes, and alteration of the voltage is sufficient to change the electrical mobility, and therefore the particle size, of the fraction selected. Automation of the system (43) enables a range of different classified monodisperse aerosols to be passed through a test filter under computer control.

The testing of filters against monodisperse particles of 1 to $> 10\ \mu$m produced by the vibrating orifice aerosol generator can be automated, but in this case it is necessary to change the aerosol-generating solution. This has been achieved (44) by apparatus in which a range of prepared solutions kept in pressurised reservoirs are connected via computer-controlled valves to a manifold from which they can be fed to the aerosol generator.

Size-selective testing with polydisperse aerosols

Obtaining the penetration curve of a filter over a range of particle sizes by using a polydisperse aerosol and a size-sensitive detector is an appealing option, for it promises a considerable amount of information from a single measurement. To some extent this promise is fulfilled, but there are difficulties in carrying out such measurements.

Technical problems with instrumentation

Technical problems with size-selectors are: their sizing process may be imperfect; they may not size particles according to the same criterion as that causing particle capture in the filter; they may see phantoms, resulting in gross mis-sizing of particles; and they will have a finite bin-width which may be unacceptably high.

Particle size can be defined in a number of ways, geometric and aerodynamic size being perhaps the two most important. The measurement of aerodynamic diameter of particles that pass through a filter acting by either inertial impaction or gravitational settling is an instance where filtration and sizing method match (45), as indeed is the measurement of particles that are filtered by diffusion, using a diffusion battery analyser (46).

Many real-time particle counters measure the optical scattering diameter (OSD) of the particle. In the case of liquid particles such as the spheres making up oil mist, optical interference effects mean that the quantity of light scattered from wavelength-sized particles does not increase monotonically with particle size. The result is that the physical diameter of a particle may not be a single-valued function of the measured scattered light (5, 47). Irregular particles are more of a problem, since particles with a wide range of geometric or aerodynamic diameters may have the same OSD; and no filter captures particles according to their optical scattering diameter. The general effect of these problems is to reduce detail in the results, so that what the experiments indicate as the characteristic penetration curve of the filter is flat and devoid of features relative to the true size-selection curve.

The most serious problem, however, is the gross mis-sizing of particles. An instrument may see phantom particles as a result of imperfect behaviour of the detection apparatus used to size the particles. A fraction of small particles may be seen by an instrument as much larger than their own physical size (48, 49). Many polydisperse aerosols have a much larger number concentration of small particles than large, and phantom particles may occur in much greater numbers than the real ones, so that the large particles are attributed a penetration characteristic of the small particles mimicking them.

The problem of finite bin-width is less severe; but it has not received much consideration, and so a brief account of it will be given here. A wide bin has the advantage that a large number of particles will be counted, resulting in a reduction in statistical error, but it will have the drawback that the particles at the bin edges will be significantly different in their degree of penetration through the filter from the particles at the centre of the bin. The effects of the first derivative of penetration against particle size will cancel out, but the second derivative will cause error. It is clear that there will be an optimal value of bin width, but the actual calculation of it will require values of numerical coefficients, and these will be system specific.

The reasoning above applies to a single bin. If measurements are made with a range of adjacent bins data reduction methods can be used to optimise the results.

Methods of production of polydisperse test aerosols

A variety of techniques can be used to produce polydisperse aerosols. The Wright dust feed (50) is a device whereby a powder compact is slowly peeled with a rotating knife. A dispenser has been developed to give economical dispersal of standard samples using a piston and knife assembly which detaches a sample of dust from a plug in a chamber where it is dispersed by compressed air (51). A method suitable for production of large quantities of airborne dust from a powder is a suction method in which a sample is prepared by allowing the dust to fill a circular groove on a rotating table. The surplus material is swept away flush with the table top to leave the groove uniformly filled with dust, which is then taken by means of a venturi (52). A popular method is that in which the dust is dispersed by means of a fluidised bed, made up either of the dust itself or of small metallic spheres used specially for the purpose, and with or without the aid of a vibrator (53, 54). The advantage of the fluidised bed is that it tends to break up agglomerates of dust because particles undergo many collisions before leaving the apparatus. Dust generated in this way tends to carry a high level of charge, and this needs to be controlled by means of a neutraliser. Polydisperse aerosols can also be produced by atomisation of a solution (18), the droplets of which lose the solvent by evaporation to give a solid aerosol, much finer than its liquid precursor; or by the combustion of specially prepared material.

Size distribution of polydisperse aerosols

In the case of polydisperse aerosols produced by making a powder airborne, the size distribution of the aerosol will be related to that of the powder from which it is produced. A number of readily available standards exist, including the dust recommended for use in clogging tests in BS2091 (55) and the ASHRAE (56) test dust (which consists of 72% by mass air cleaner fine test dust, 23% molocco black and 5% cotton linters). Commercial abrasive dusts, which are produced in a range of grades, can be used for a variety of aerosol studies, including filter testing (57, 58). With non-standard powders, typical values of Geometric Standard deviation are observed to be 1.8–2.7 for isometric particles (53), and with fibrous particles (59) similar values are obtained for the variation of both length and diameter.

The size distribution of an aerosol produced by atomisation, the BS4400 sodium chloride aerosol, is given in the form of a cumulative mass distribution in Fig. 9.5. If the aerosol were lognormally distributed the cumulative distribution of the plot shown would be a straight line. The deviation is attributed to the removal of large particles by impaction on baffles in the atomiser.

The size distribution of atomised aerosols has been measured, by electrostatic classification (60) as a function of atomiser pressure and solute concentration. Both the geometric number mean particle diameter and the geometric standard deviation (GSD) increase with the concentration of the aerosol-generating solution. The number mean does not increase as the cube root of the concentration, the relationship that would be expected only if the GSD of the spray did not alter. Both parameters decrease with increasing atomiser pressure, an increase in the latter from 300 to 600 kPa resulting in a decrease in number geometric mean diameter from 0.05 to 0.04 μm.

Filter testing by sample collection and analysis

An alternative to real time testing is the collection of aerosol samples before and after filtration, followed by analysis of the collected

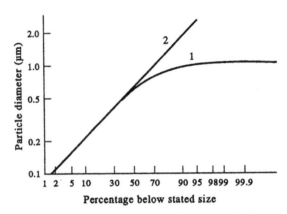

FIG. 9.5 Log-probability plot of the cumulative mass distribution of BS4400 sodium chloride aerosol: (1) aerosol with large particles removed; (2) lognormally distributed aerosol with the same median particle size. (Reproduced from BS4400 (1969) by permission of the British Standards Institution. Complete copies of the standard may be obtained by post from BSI Publishers, Linford Wood, Milton Keynes, MK14 6LE.)

material. Size-selective analysis can be carried out by microscopy (61) and this is particularly suitable if the aerosols are fibrous, and, therefore, difficult to size in real time. Alternatively size analysis can be carried out by Coulter Analysis after dispersion. Certain aerosols have a tendency to agglomerate, and dispersion can separate particles that have travelled through the filter together. In such an instance the resultant small particles may have attributed to them behaviour truly characteristic of larger ones.

Non size selective testing with polydisperse aerosols

Non size-selective testing of filters with polydisperse aerosols gives the least information of all test methods, but it is of special importance because many standard test methods use the process. The behaviour of polydisperse aerosols has been described in Chapter 1, the penetration being given by,

$$P(h) = \int_0^\infty A(\alpha)\exp(-\alpha h)\, d\alpha \qquad (9.5)$$

where $A(\alpha)$ is a weighting function describing the relative frequency of particles with layer efficiencies α. As shown in Fig. 1.5, the penetration curve is concave upwards, since the aerosol becomes more penetrating as it passes through the filter. A functional form could be ascribed to

$A(\alpha)$ if it were possible thoroughly to analyse the aerosol and filtration mechanisms, but some understanding of the form that the function should take follows from simple arguments. It has been shown that a most-penetrating particle exists, which means that some lower bound can be placed on α, below which $A(\alpha)$ vanishes. The existence of this most penetrating particle means that there must be a spread of sizes close to it where the penetration varies relatively little, which means in turn that $A(\alpha)$ in this region rises relatively quickly. Very large particles are captured with increasing efficiency, and so for large values of α we would expect the function to reach zero gradually. The simplest type of function that satisfies the above requirements is a positively skewed unimodal function. The lognormal and Weibull functions behave in this way and, since the latter can have a cut-off at non-zero α, it may fit the data; but it has been found in practice (62) that the behaviour of polydisperse test aerosols is particularly well-described by the gamma distribution.

$$A(\alpha) = \frac{\exp(-\alpha\theta_g)\alpha^{\gamma_g - 1}\theta^{\gamma_g}}{\Gamma(\gamma_g)} \qquad \begin{array}{l} \alpha \geq 0 \\ \theta_g, \gamma_g > 0 \end{array} \qquad (9.6)$$

Substitution of this function into equation 9.5 results in a particularly simple integral, which gives,

$$P(h) = \left(\frac{\theta_g}{\theta_g + h}\right)^{\gamma_g}. \qquad (9.7)$$

A function of this form, with two unknown parameters, can be fitted to the results of experimental data, and an example of such fits to experimental points is given in Fig. 9.6.

Aerosol number distributions and mass distributions

The statistics of aerosol size distributions must be considered in the interpretation of test results because in addition to actual numbers of particles challenging the filter or penetrating through it, a weighting function must be included, the nature of which depends on the particular aerosol property that is being measured by the detector. Two particles of different size will be considered equivalent so far as a number count is concerned, but if the total mass of aerosol is of interest the particles must be weighted according to their mass. If the normalised particle size distribution of an aerosol is given by $S(d_p)$, i.e.

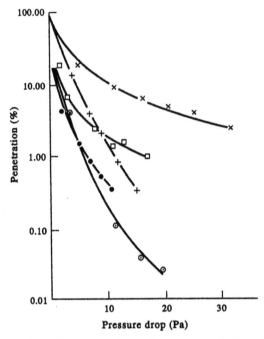

Fɪɢ. 9.6 Penetration of a standard polydisperse aerosol through a range of homogeneous filters of various thicknesses (62). (© Crown Copyright; by permission of Her Majesty's Stationery Office.)

if the number of particles with diameters around d_p is proportional to $S(d_p)$, then the number mean particle diameter, μ_n, will be,

$$\mu_n = \int_0^\infty S(d_p)\, d_p\, d(d_p) \tag{9.8}$$

and the mass (or volume) mean diameter, μ_m, will be given by,

$$\mu_m = \frac{\displaystyle\int_0^\infty S(d_p)\, d_p^4\, d(d_p)}{\displaystyle\int_0^\infty S(d_p)\, d_p^3\, d(d_p)}. \tag{9.9}$$

The other statistical parameters are weighted in corresponding ways. If an aerosol is monodisperse all of the means equal the actual size since S will then be a delta function; and the difference between them increases with the degree of poly-dispersity.

Size distributions are frequently unimodal and positively skewed, and they are often approximated by lognormal distributions (63). True lognormal distributions appear as straight lines when the cumulative distributions are plotted on log-probability plots, and often the straightness of a plot made with experimental data is used as a measure of the extent to which the distribution resembles a lognormal. In any case plotting data in this way enables medians and percentiles to be read with ease whatever the nature of the distribution.

Figure 9.7 shows a log-probability plot and Graph 1 is the cumulative mass distribution of a DOP aerosol with a mass median or geometric mean (the two are identical for a lognormal distribution) size of 1.65

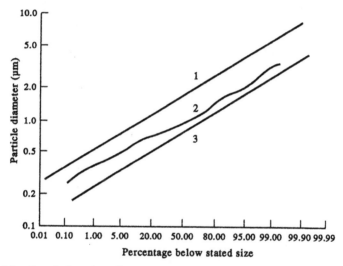

FIG. 9.7 Cumulative size distributions on a log-probability plot: (1) mass distribution of DOP aerosol with mass median diameter of 1.65 μm; (2) size distribution of same aerosol measured by light scattering (64) (by permission of the Air and Waste Management Association); (3) number distribution of the aerosol calculated using equation 9.10.

μm and a geometric standard deviation (GSD) of 0.53 (64); Graph 3 is the calculated cumulative number distribution of the same aerosol, which has a median of 0.75 μm, and the same geometric standard deviation. The number median, d_n, and the mass median d_m, are related by,

$$\bar{d}_m = \bar{d}_n + 3\sigma_g^2 \tag{9.10}$$

and equation 9.10 shows that for a given number median the mass median increases with the GSD, σ_g. The difference between the two medians means that although 80% of the number of particles in the aerosol are smaller than 1.4 μm in diameter, 80% of the mass is contained in particles that are larger. Graph 2 of Fig. 9.7 shows a cumulative distribution of the aerosol measured experimentally by light scattering. The example illustrates that a polydisperse aerosol can be seen in different ways depending on which of its properties (number, mass or optical scattering diameter in this case) is being considered.

Effect of detection method on ascribed penetration

The significance of the particle statistics discussed above is that the detection method affects the value recorded for aerosol penetration. The behaviour of an aerosol is, for thin filters, similar to that of a monodisperse aerosol with the same mean layer efficiency, but the weighting of this mean will depend on the detection method. If the aerosol is detected by a method that depends on the mass present, like the flame photometry used in the BS4400 aerosol test, the appropriate mean will be the mass mean (which is 0.6 μm). If the detection method is light scattering something like the (smaller) surface area mean will be appropriate, and if the particles are detected by counting the number mean (approximately 0.06 μm in the specific case cited) must be used. Which of these mean-sized particles is the most penetrating depends on the mechanism of particle capture, but it is very likely that they will be different, as will the results of the experiments.

A combined theoretical and experimental approach (65, 66) compared the response of a condensation nucleus counter, which counts aerosol particles, and an aerosol photometer, which responds to a function of size roughly similar to surface area. Both instruments were used to measure aerosol penetration in the same aerosol/filter combination, but the two devices recorded up to an order of magnitude difference in the aerosol penetration. The particle size distribution seen by the CNC has its mode closer to the number mean; that seen by the photometer has its mode closer to the surface area mean. The higher penetration is recorded by the instrument that sees a mode closer to the size of the most penetrating particle, which means that the CNC will tend to record the higher penetration for coarser aerosols and the photometer for finer. The discrepancy increases with the GSD of the size distribution since this is related to the difference between the two modes.

The responses of the two instruments are similar in the case of aerosol that does not penetrate through the filter but which leaks, especially if the leaks are large. In this situation there is no most-penetrating particle size, since the aerosol size distribution is not significantly altered during leakage (67). In a separate exercise, a study of the penetration through a filter of monodisperse aerosols (generated both by an electrostatic classifier and by a vibrating orifice generator) gave results that were insensitive to the detection method used (68), since in the case of a monodisperse aerosol the two modes are the same.

Testing of high efficiency filters

Filters of very high efficiency present special problems, because of the enormous difference in concentration between the filtered and the unfiltered aerosols. The former may be made high in order to prevent the latter from being too low, and the high concentration may encourage coagulation of the aerosol (69) or reduce the lifetime of the filter by clogging (70). It may affect the performance of the detection instrument by causing coincidence errors (71), or it may even saturate the instrument's response. Both problems may be avoided or at least reduced by the use of a diluter (72). Such devices work best when they are used against small particles, because these are efficiently removed by the filter incorporated, but not by the by-pass line. It is possible to obtain dilution of 3000:1 in a single stage (73), but the observed dilution may be more variable than the nominal (5).

The problem of low concentration of filtered aerosols is alleviated by the development of analysis techniques that are extremely sensitive, and a review of these (69) mentioned chemical analysis such as flame photometry, in which the particles are passed into a clean hydrogen flame, and the characteristic light emitted is analysed. In addition radioactively labelled aerosols or fluorescent aerosols may be used (74).

Small particles cause a particular detection problem, because mass estimation is extremely difficult even if they exist in large numbers. Optical scattering methods are difficult to use because the amount of light scattered by small particles varies as the sixth power of the radius. In general white light instruments have a lower size limit of about 0.5 μm and laser-based instruments one of 0.1 μm. Low noise devices can extend the lower size-detection limit, as may the suggested use of the open cavity of a laser (75).

The estimation is made easier by the use of a condensation nucleus counter (43), in which the particles are treated as condensation nuclei in a process like aerosol production from a Lamer Sinclair generator. The particles grow, by the accretion of volatile matter, to such an extent that they can be readily seen by conventional detection methods like optical scattering or, if their concentration is low, by counting techniques (76). If the aerosol particles are detected by counting, filtration efficiency may be so high that only a few particle counts per hour are observed. At this low rate it is important that real counts are distinguishable from spurious effects attributable to the behaviour of the detection instrument.

Tests with electrically charged aerosols

Since electrical effects can be critical in filtration it may be useful to carry out tests on filters with electrically charged aerosols or with aerosols that are known to be completely neutral. One method of producing charged submicrometre aerosols is the electrostatic classification method described above, but without a neutraliser to remove the charge (77). An electric charge can be placed on larger particles, produced from a vibrating orifice aerosol generator, by using an induction method (78) in which a high voltage electrode is placed immediately above the orifice where the aerosol is produced.

Where electrical effects are significant, only an aerosol in which all the particles carry the same electric charge will behave as a monodisperse aerosol (77). Aerosols with single size but a range of charges will behave as if they were polydisperse. A measurement method analogous to testing with polydisperse aerosols is that of employing a monodisperse aerosol with a range of charges, subjected to mobility analysis both before and after being passed through the test filter (79, 80). Experiments of this sort could, in principle, be extended to involve a polydisperse test aerosol with a range of charges, subjected to both size analysis and electrical mobility analysis; though the large number of channels needed to accommodate the data would mean that protracted experiments would be needed in order to make channel counts significant.

Neutralisation of aerosols

As mentioned above, the neutralisation of aerosols is frequently necessary. The most popular type of neutralising device is the radioactive ionising neutraliser (81), which consists of a small source, usually a few mCi of the low toxicity isotope, Krypton 85. The term "neutralisation" is not strictly correct because the aerosols invariably have a residual charge, but it is sanctioned by idiom. "Neutralised" aerosols with their low level of charge (usually slightly higher than the equilibrium level), must be distinguished from truly neutral aerosols, which are completely uncharged.

The use of neutralisers presents no serious technical problems if all that is required is the reduction of troublesome aerosol charge to a manageable level. If, however, the actual charge level on the aerosols is critical, the time-dependent nature of the action of a neutraliser may need to be considered. Most of the ions produced by such devices are removed by recombination; and since the ionisation of neutral atoms or molecules involves the production of equal numbers of positive and negative ions, complete recombination is possible. The value of the recombination coefficient, Rc, for small ions is typically 1.6×10^{-6} $cm^3 \, s^{-1}$ (82), and the recombination equation, giving the number density of ions after time t, is,

$$N_1(t) = \frac{N_1(o)}{1 + R_c N_1(o)t}. \tag{9.11}$$

A significant number of ions will persist for at least thirty seconds after the air has left the neutraliser, as shown in Fig. 9.8, which depicts experimental measurements of the ionic concentration in air leaving such a device (83). This means first that neutralisation is possible without passing the aerosol through the neutraliser but simply by mixing it with ionised air, and second that the process of neutralisation continues after the aerosol has left the neutraliser itself. The criterion for effective neutralisation can be expressed in terms of the product of the ionic concentration, N_1, and the exposure time, t, a typical value of 10^5 ions/ml.sec being needed for the neutralisation of small particles (81).

An alternative system, which has the advantage of special applicability to high concentration aerosols, and which does not have the handling problems associated with radioactive sources, is a bipolar corona device (84, 85), which is capable of dealing with aerosol concentrations of up to 200 mg/m^3.

A direct indication of the effect of neutralisation on filtration efficiency has been shown on sodium chloride aerosols (86) and on aerosols of lead fume and silica (87). The filters selected appear to include both electrostatic and mechanical types. No quantitative measure of the charge on the aerosol was made either before or after it had been passed through the neutraliser, but in most instances a significant increase in penetration was observed as a result of whatever reduction in charge had taken place. A similar effect has been observed by other workers (84).

Worst possible case testing

It has been shown in previous sections that there exists a particle size at which aerosol penetration is greatest. The exact size depends on the filter fibre size and on the filtration velocity, but it is usually submicrometre. If a filter is tested against this type of particle then one

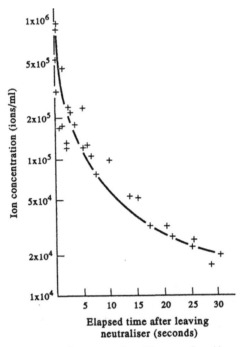

FIG. 9.8   Time-dependence of concentration of ions produced by radioactive aerosol
neutraliser as a function of time: + experiment; — theory.

can be confident that the filter will perform at least as well, and possibly
considerably better, against other aerosols. The approach of testing
with such aerosols is most suitable to the situation where extremely
high efficiency is required from a filter that normally acts against
particles of sizes close to the most penetrating, such as a clean room
filter. The result will be a conservative specification, with the possible
result that the working aerosol penetrates by an order of magnitude less
than the test aerosol (40). In other situations it may result in a
considerable over-specification of performance, so that a filter that
gives good initial performance may have a resistance that is
unacceptably high, or may be more susceptible to clogging than a filter
with a more realistic specification.

The worst case aerosol can be selected with the apparatus described
above (43) producing electrostatically classified monodisperse aerosols
(88); and a comparison has been made between these and more
conventionally generated aerosols of lead fume and silica (87). The
most penetrating size was usually found to be in the region of 0.1 μm in
diameter, agreeing with the result presented in Chapter 4. The lead
fume and silica aerosols were invariably less penetrating than the

submicrometre monodisperse aerosols. A similar approach has been used by other workers on essentially the same system (89).

Realistic prediction of filter performance

What is most desirable from a test of filter performance is a realistic indication of how well the filter will perform when subjected to real working conditions; but the use of real aerosols in tests causes problems. The composition of atmospheric aerosols from a variety of sources will change from day to day and so will the extent of their penetration through filters (6, 90). Industrial aerosol concentration may peak in such a way as to cause coincidence errors in the instrumentation (91). Laboratory simulation using aerosols of similar chemical composition to the real aerosols is also problemational, because the physical state of an aerosol is usually more important in fixing its capture efficiency than is its chemical composition; and features like particle size, particle charge and extent of agglomeration of the laboratory aerosol may be quite different from that of the industrial aerosol that it is meant to simulate.

It would be useful to have an estimate not only of the initial performance of a filter but also of the way in which the filter is affected by loading, leading to an estimate of the lifetime of the filter. Although in principle it is an easy matter to load a filter with dust, the procedure has all of the problems of real tests outlined above along with the very serious problem discussed in Chapter 8, that although ultimate clogging is inevitable, the point at which it occurs depends critically upon the rate at which aerosol loading is carried out.

Standard filters

For comparisons among test methods themselves, it is useful to have standard filters. The preparation of filters for a comparison between two filter test instruments in the same laboratory has been attempted (92), the filters in question being in the form of one or many layers of commercially available glass-fibre material of different grades or melt-blown polymer fibre filters (73). The filters were constructed in such a way as to give penetrations that differed successively by an order of magnitude.

There are no generally-accepted standard filters, but other candidates are the model filters described in Chapter 2. The requirements of standard filters are: that they should be available in a wide range of efficiencies, that they should be consistent and reliable in their behaviour, and that their performance should alter very little as they become loaded with aerosol.

References

1. FAIN, D. E. Standards for pressure drop testing of filters as applied to HEPA filters. In *Fluid Filtration: Gas* (ed. R. R. RABER), Volume 1, ASTM STP 975, pp. 364–379. American Society for Testing and Materials, Philadelphia, 1986.

2. HEIDENREID, E., TITTEL, R., NEUBER, A. and ADAM, R. Measuring the inhomogeneities of high efficient glass fibre filter media. *J. Aerosol Sci.*, 1991, **22**, Supp 1, S785–S788.

3. DORMAN, R. G. European and American methods of testing air conditioning filters. *Filtr. Sep.*, Jan./Feb. 1968, 1–5.

4. RIVERS, R. D. and MURPHY, D. J. Air filter testing: Current status and future prospects. In *Fluid Filtration: Gas* (ed. R. R. RABER), Volume 1, ASTM STP 975, pp. 214–237. American Society for Testing and Materials, Philadelphia, 1986.

5. GROSS, H. New testing procedure for standard absolute (HEPA) and high performance absolute (ULPA) filters. *10th International Symposium on Contamination Conrol, Zurich*, 10–14 *Sept.*, 1990, 292–264.

6. BAUER, E. J., REAGOR, B. T. and RUSSELL, C. A. Use of particle counts for filter evaluation. *ASHRAE J.*, 1973, 53–59.

7. DAVIES, C. N. *Dust is Dangerous*, Faber and Faber, 1954.

8. MATSUI, H., IKEZAWA, Y., YOSHIDA, Y., YOKOCHI, A., MATSUMOTO, S., SUGITA, N. and MIKAMI, S. Mixing of duct air for the representative air sampling in *in situ* testing of high-efficiency particulate air filters. *Proceedings of the 3rd International Air Conference*, 24–27 *Sept.* 1990, *Kyoto*, Pergamon, 1990.

9. LEARY, A. G. and RADDEN, E. B. Evaluate performance of HEPA filters on clean benches. *Heat. Piping/Air Cond.*, Sept. 1967, 117–119.

10. FAHRBACH, J. The effects of leaks on total penetration velocity and concentration measurements at perforated filters. *Staub Reinhalt. Luft.*, 1970, **30**, 45–52 (in English).

11. BLACKFORD, D. B., HANSON, A. E., KINNEY, P. and ANANTH, G. P. Details of recent work towards improving the performance of the TSI aerodynamic particle sizer. *Aerosol Society Proceedings*, 1988, 311–315.

12. BLACKFORD, D. B., BROWN, R. C. and WAKE, D. A semi-automated system for measuring the penetration of monodisperse aerosols through filters. *J. Aerosol Sci.*, 1985, **16**, 415–425.

13. WAKE, D., GRAY, R. and BROWN, R. C. Assessment of vacuum cleaners of both wet and dry types for use in potteries. *Ann. Occ. Hyg.*, 1988, **32**, 201–221.

14. CLARK, M. R., TENNAL, K. B., RIMMIER, T. W. and MAZUMDER, M. K. Evaluation of particulate air filters for use in indoor air cleaning. *A.A.A.R. Conference*, 1991.

15. BROWN, K. E., BEYER, J. and GENTRY, J. W. Calibration and design of diffusion batteries for ultrafine aerosols. *J. Aerosol Sci.*, 1984, **15**(2), 133–145.

16. SCHURMANN, G. The use of particle counters and other related components for today's filter testing. The component approach for higher flexibility. In *Advances in Filtration and Separation Technology* (ed. K. L. RUBOW), Vol. 4, pp. 68–71. American Filtration Society, 1991.

17. BLACKFORD, D. B. and HARRIS, G. W. Field experience with SIMSLIN II—a continuously recording dust sampling instrument. *Ann. Occ. Hyg.*, 1978, **21**, 301–313.

18. British Standard BS4400. Method for sodium chloride particulate test for respirator filters. *British Standards Institution, London*, 1969.

19. BARON, P. A. Calibration and use of the aerodynamic particle sizer (APS 3300). *Aerosol Sci. Technol.*, 1986, **5**(1), 55–67.

20. FUCHS, N. A. and SUTUGIN, A. G. Generation and use of monodisperse aerosols. In *Aerosol Science* (ed. C. N. DAVIES), Academic Press, New York, 1966.

21. MAY, K. R. An improved spinning top homogeneous spray apparatus. *J. Appl. Phys.*, 1949, **20**, 932–938.

22. BERGLUND, R. N. and LIU, B. Y. H. Generation of monodisperse aerosol standards. *Environ. Sci. Tech.*, 1973, **7**, 147–153.

23. RAYLEIGH. On the equilibrium of liquid conducting masses charged with electricity. *Philos. Mag.*, 1882, **14** Series 5, 184–186.

24. WHITBY, K. T. and LIU, B. Y. H. The electrical behaviour of aerosols. In *Aerosol Science* (ed. C. N. DAVIES), Academic Press, New York, 1966.

25. Leong, K. H. Morphology of aerosol particles generated by the evaporation of solid drops. *J. Aerosol Sci.*, 1981, **12**(5), 417–435.

26. Sinclair, D. and Lamer, V. Light scattering as a measure of particle size in aerosols. The production of monodisperse aerosols. *Chem. Rev.*, 1949, **44**, 245–267.

27. Whitby, K. T. and Liu, B. Y. H. Polydisperse aerosols. Electrical charge and residual size distribution. *Atmos. Environ.*, 1968, **2**, 103–116.

28. Raabe, O. G. The dilution of monodisperse suspensions for aerosolization. *Ind. Hyg. Assoc. J.*, 1968 Sept./Oct., 439–443.

29. Liu, B. Y. H. and Pui, D. Y. H. A submicron aerosol standard and the primary absolute calibration of the condensation nucleus counter. *J. Coll. Int. Sci.*, 1974, **47**(1), 155–171.

30. Knutson, E. O. and Whitby, K. T. Aerosol classification by electrical mobility: Apparatus, theory and applications. *J. Aerosol Sci.*, 1975, **6**, 443–451.

31. Spurny, K. R. Preparation of size-selected fibers and fibrous aerosols for biological experiments. *Environ. Int.*, 1980, **4**, 39–46.

32. Ogden, T. L. and Walton, W. H. The descent angle of inclined cylinders. A possible means of separating fibres by length. *Ann. Occ. Hyg.*, 1975, **18**, 157–160.

33. Esmen, N. A., Kahn, R. A., LaPietra, D. and McGovern, E. D. Generation of monodisperse fibrous glass aerosols. *Am. Ind. Hyg. Assoc. J.*, 1980, **41**, 175–179.

34. Loo, B. W., Cork, C. P. and Madden, N. W. A laser based monodisperse fiber generator. *J. Aerosol Sci.*, 1982, **13**, 241–248.

35. Zebel, G., Hochrainer, D. and Boose, C. A sampling method with separated deposition of airborne fibers and other particles. *J. Aerosol Sci.*, 1977, **8**, 203–213.

36. Zebel, G. and Hochrainer, D. A simple device with separation of fibres and isometric particles. *J. Aerosol Sci.*, 1979, **10**(2), 245.

37. Griffiths, W. D. The shape-selective sampling of fibrous aerosols. *J. Aerosol Sci.*, 1988, **19**(6), 703–713.

38. Vaughan, N. P. The generation of monodisperse fibres of caffeine. *J. Aerosol Sci.*, 1990, **21**(3), 453–462.

39. Hoover, M. D., Casalnuovo, S. A., Lipowicz, P. J., Yeh, C. Y., Hanson, R. W. and Hurd, A. J. A method for producing non-spherical monodisperse particles using integrated circuit fabrication techniques. *J. Aerosol Sci.*, 1990, **21**(4), 569–575.

40. Scripsick, R. C. New filter efficiency tests being developed for the DOE. In *Fluid Filtration: Gas* (ed. R. R. Raber), Volume 1, ASTM STP 975, pp. 345–363. American Society for Testing and Materials, Philadelphia, 1986.

41. Flannery, J. L. and Walcroft, J. P. Air cleanliness validation for cleanrooms. In *Fluid Filtration: Gas* (ed. R. R. Raber), Volume 1, ASTM STP 975, pp. 390–401. American Society for Testing and Materials, Philadelphia, 1986.

42. Wake, D. and Brown, R. C. Filtration of monodisperse aerosols and polydisperse dusts by porous foam filters. *J. Aerosol Sci.*, 1991, **22**(6), 693–706.

43. Remiarz, R. J., Agarwal, J. K., Nelson, P. A. and Moyer, E. A new automated method for testing particulate filters. *J. ISRP*, 1984, **2**(3), 275–287.

44. Wake, D. An automated system for measuring the penetration of respirable-sized aerosols through filters. *European Aerosol Society Conference*, 1992.

45. Wake, D. and Brown, R. C. Testing of filters using aerosols with a single particle size and a range of sizes. *First Aerosol Conference Proceedings*, 1987, 49–52.

46. Sinclair, D. Penetration of HEPA filters by submicron aerosols. *J. Aerosol Sci.*, 1976, **7**, 175–179.

47. Gebhart, J., Blankenberg, P., Borman, S. and Roth, C. Counting efficiency and sizing characteristics of optical particle counters. *J. Aerosol. Sci.*, 1984, **15**, 279–281.

48. Wake, D. Anomolous effects in filter penetration measurements using the aerodynamic particle sizer (APS 3300). *J. Aerosol Sci.*, 1989, **20**(1), 13–17.

49. Heitbrink, W. A. and Baron, P. A. An approach to evaluating and correcting aerodynamic particle sizer measurements for phantom particle creation. *Am. Ind. Hyg. Assoc. J.*, 1992, **53**(7), 427–431.

50. Wright, B. M. A new dust feed mechanism. *J. Sci. Instrum.*, 1950, **27**, 12–15.

51. Timbrell, V., Hyett, A. W. and Skidmore, J. W. A simple dispenser for generating dust clouds from standard reference samples of asbestos. *Ann. Occ. Hyg.*, 1968, **11**, 273–281.

52. HATTERSLEY, R., MAGUIRE, B. A. and TYE, D. L. A laboratory dust cloud producer. *SMRE Research Report No.* 103, 1954.

53. MARPLE, V. A., LIU, B. Y. H. and RUBOW, K. L. A dust generator for laboratory use. *Am. Ind. Hyg. Assoc. J.*, 1978, **39**, 26–32.

54. SPURNY, K. R., WEISS, G. and OFIELA, H. Zur Herstellung von Feinstaubproben fur biologische Unterschungen. *Staub Reinhalt. Luft.*, 1978, **38**(10), 417–420.

55. BRITISH STANDARD BS2091. Specification of respirators for protection against harmful dusts, gases and scheduled agricultural chemicals. *British Standards Institution*, London, 1969.

56. ASHRAE 52 76. Method of testing air cleaning devices used in general ventilation air for removing particulate matter. *American Society of Heating Refrigeration and Ventilation Engineers, Atlanta, GA*, 1976.

57. MARK, D., VINCENT, J. H., GIBSON, H. and WITHERSPOON, W. A. Applications of closely graded powders of fused alumina as test dusts for aerosol studies. *J. Aerosol Sci.*, 1985, **16**, 125–131.

58. BRITISH STANDARD BS2831. Method of testing of air filters used in air conditioning and general ventilation. *British Standards Institution*, London, 1971.

59. CARPENTER, R. L., PICKRELL, J. A., MOKLER, B. V., YEH, H. C. and DeNEE, P. B. Generation of respirable glass fiber aerosols using a fluidized bed generator. *Am. Ind. Hyg. Assoc. J.*, 1981, **42**(11), 777–784.

60. McDERMOTT, W. T., SHWARZ, A. and THOROGOOD, R. M. Large scale filter testing for high purity gas supply systems used in semiconductor processing. *Aerosol Sci. Technol.*, 1988, **9**, 1–14.

61. SPURNY, K. R. Aerosol filter testing and filtration studies by electron microscopical methods. *5th World Filtration Congress Proceedings*, 632–634.

62. BROWN, R. C., DAVIES, J. K. W. and WAKE, D. Penetration of test aerosols through filters described in terms of a gamma distribution of layer efficiencies. *J. Aerosol Sci.*, 1987, **18**, 499–509.

63. AITCHISON, J. and BROWN, J. A. C. *The Lognormal Distribution*, CUP, Cambridge, 1963.

64. FIRST, M. W. and HINDS, W. C. High velocity filtration of submicrometer aerosols. *J. Air. Pollut. Control Assoc.*, 1976, **26**(2), 119–123.

65. BERGMAN, W. and BIERMANN, A. Effect of DOP heterodispersion on HEPA filter penetration measurement. In *Aerosols* (ed. B. Y. H. LIU, D. PUI and H. FISSAN), Elsevier, 1984.

66. BIERMANN, A. H. and BERGMAN, W. Filter penetration measurements using a condensation nuclei counter and an aerosol photometer. *J. Aerosol Sci.*, 1988, **19**(4), 471–483.

67. HINDS, W. C. and KRASKE, G. Performance of dust respirators with facial seal leaks, 1. Experimental. *Am. Ind. Hyg. Assoc. J.*, 1987, **48**, 842–847.

68. FARDI, B. and LIU, B. Y. H. Performance of disposable respirators. *Part. Syst. Charact.*, 1991, **8**, 308–314.

69. SPURNY, K. Measuring the separation efficiency of high-efficiency filters. *Staub Reinhalt. Luft.*, 1970, **30**(12), 40–45 (in English).

70. PARKER, R. C., MARSHALL, M. and BOSLEY, R. B. A new method for *in-situ* filter testing using pulses of aerosol. *European Aerosol Conference*, 1992.

71. ORTIZ, J. F. Preliminary test results of a round robin test program to evaluate a multi-state HEPA filter system using single particle size counters. In *Advances in Filtration and Separation Technology* (ed. K. L. RUBOW), Vol. 4, pp. 213–216. American Filtration Society, 1991.

72. MOELTER, W. Fast automated testing of HEPA and ULPA filters with a new filter test unit. Particle technology in relation to filtration and separation. *Royal Flemish Society of Engineers*, 1988, 4.109–4.113.

73. NICHOLSON, R. M. An update on current and developing test standards for air/gas filtration in ASTM F21.20. In *Advances in Filtration and Separation Technology* (ed. K. L. RUBOW), Vol. 4, pp. 207–212. American Filtration Society, 1991.

74. MULCEY, P., PYBOT, P. and VENDEL, J. Real time detection of a fluorescent aerosol. Application to the efficiency measurements of HEPA filters. *5th World Filtration Congress Proceedings*, 635–640, 1986.

75. KNOLLENBERG, R. G. Filter evaluation using particle size spectrometers. *Ann. Tech. Meet. Proc. Inst. Environ. Sci.*, 1982, 171–175.

76. KING, J. G. Air cleanliness requirements for clean rooms. In *Fluid Filtration: Gas* (ed. R. R. RABER), Volume 1, ASTM STP 975, pp. 383–389. American Society for Testing and Materials, Philadelphia, 1986.

77. TROTTIER, R. A. and BROWN, R. C. The effect of aerosol charge and filter charge on the filtration of submicrometre aerosols. *J. Aerosol Sci.*, 1990, **21** Supp. 1, S689–S692.

78. REISCHL, G. P., JOHN, W. and DEVOR, W. Uniform charging of monodisperse aerosols. *J. Aerosol Sci.*, 1977, **8**, 55–66.

79. JOHNSTON, A. M. A semi-automated method for the measurement of charge carried by airborne dust. *J. Aerosol Sci.*, 1983, **14**, 643–655.

80. WAKE, D., THORPE, A., BOSTOCK, G. J., DAVIES, J. K. W. and BROWN, R. C. Apparatus for measurement of the electrical mobility of aerosol particles: computer control and data analysis. *J. Aerosol Sci.*, 1991, **22**(7), 901–916.

81. LIU, B. Y. H. and PUI, D. Y. H. Electrical neutralization of aerosols. *J. Aerosol Sci.*, 1974, **5**, 465–472.

82. CHARRY, J. M. and KAVET, R. I. *Air Ions: Physical and Biological Aspects*, CRC Press, Florida, 1987.

83. THORPE, A. and BROWN, R. C. Charging and neutralization of aerosols, to be published.

84. ADACHI, M., LIU, B. Y. H. and PUI, D. Y. H. Development of a corona neutralizer. *Annual Meeting of the A.A.A.R.*, 1989, 320.

85. ZHANG, Z. Q., JOHNSON, B. and AGARWAL, J. Aerosol neutralization for filter testing. In *Advances in Filtration and Separation Technology* (ed. K. L. RUBOW), Vol. 4, pp. 72–75. American Filtration Society, 1991.

86. REED, L. D., SMITH, D. L. and MOYER, E. S. Comparison of respirator particulate filter test methods. *J. ISRP*, 1986, **4**(3), 43–60.

87. MOYER, E. S. and STEVENS, G. A. Worst case aerosol testing parameters: III. Initial penetration of charged and neutralized lead fume and silica dust aerosols through clean, unloaded respirator filters. *Am. Ind. Hyg. Assoc. J.*, 1989, **50**(5), 271–274.

88. STEVENS, G. A. and MOYER, E. S. Worst case aerosol testing parameters: I. Sodium chloride and dioctyl phthalate aerosol filter efficiency as a function of particle size and flow rate. *Am. Ind. Hyg. Assoc. J.*, 1989, **50**(5), 257–264.

89. NIEMINEN, K. and PUUTIO, M. An alternative test system for respirator filters. *Proceedings of the 3rd International Aerosol Conference, Kyoto*, 24–27 Sept. 1990, 783–785.

90. RABER, R. R. Development of an artificial dust spot technique for use with ASHRAE 52-76. In *Fluid Filtration: Gas* (ed. R. R. RABER), Volume 1, ASTM STP 975, pp. 229–237. American Society for Testing and Materials, Philadelphia, 1986.

91. DELACRUZ, O. M. and RABER, R. R. Experience with fractional efficiency testing of HIVAC filters. In *Advances in Filtration and Separation Technology* (ed. K. L. RUBOW), Vol. 4, pp. 464–469. American Filtration Society, 1991.

92. SIMPSON, K. L. A comparison of particulate filter tests. In *Advances in Filtration and Separation Technology* (ed. K. L. RUBOW), Vol. 4, pp. 149–152. American Filtration Society, 1991.

Conclusion

In the final chapter, the study of filtration returned to the macroscopic level described in Chapter 1. In a sense the work has turned full circle, though no book on a living subject can ever be considered complete.

A long time has passed since air filtration changed from empiricism to science; and an accepted body of knowledge is now put to good use in filter design and specification. As the knowledge increases so will both the number of situations in which filtration is used and the efficiency with which filters are employed.

There is scope for research and development in all parts of the subject, but a few parts in which it would be particularly welcome are worthy of mention. Filter structure has received less attention than its fundamental position would merit; and the development of theoretical models to deal with filters of non-uniform structure could lead to much increased understanding.

Most of the theory and experiment on filtration has dealt with aerosols made from spherical particles of low and ill-defined charge. This is a good point to start but not to finish, and more work is needed with particles of non-spherical shape and measured electric charge, particularly in the situation where the filter is electrically charged as well.

The accumulation of data from filters used in real working conditions would be valuable, even though the analysis of field data is always difficult.

It is usually possible to specify the initial performance of a filter, but as the dust load increases our understanding diminishes. A considerable amount of work has already been done on filter loading, but the subject is a difficult one, and still more effort is needed, particularly in the study of composite filters.

Some time may elapse before we have a satisfactory description of non-uniform filters loaded with electrically charged fibres; but a systematic and painstaking study will undoubtedly lead to improvements in the understanding and application of air filtration.

Index

Printed and bound by CPI Group (UK) Ltd, Croydon, CR0 4YY

03/10/2024

01040418-0017